The Geography of Black Pepper (*Piper nigrum*)

Kodoth Prabhakaran Nair

The Geography of Black Pepper (*Piper nigrum*)

The "King" of Spices – Volume 1

Springer

Kodoth Prabhakaran Nair
Villament G3
C/o Dr. Mavila Pankajakshy
Malaparamba, Kozhikode, Kerala, India

ISBN 978-3-030-52867-6 ISBN 978-3-030-52865-2 (eBook)
https://doi.org/10.1007/978-3-030-52865-2

This Springer imprint is published by the registered company Springer Nature Switzerland AG.
The registered company address is: Gewerbestrasse 11, 6330 Cham, Switzerland

To Pankajam, my wife, who is all to me

Preface

This book is about the world's most important spice crop, black pepper, known as the "King" of spices.

Black pepper has a checkered history, dating back to the times of Queen Sheeba and King Solomon (B.C. 1015–B.C. 66), which has influenced the destiny of nations and their people across the world, both economically and culturally. Today, black pepper commands the leading position among different spices as one of immense commercial importance in world trade, and is finding its way onto dining tables of millions around the world, even in Europe, North America, and Japan. Pepper use ranges from a simple dietary constituent to that of immense pharmacological benefit. Though beset with many problems, both agronomic and economic, it is a safe bet that pepper will emerge as the world's most sought-after spice; its global demand is predicted to escalate colossally to about 280,000 metric tons by the year 2020, which will further climb to 360,000 metric tons by 2050.

This book's exhaustive review details the various constraints in enhancing pepper productivity and charts contours of a new course of action. Among the primary constraints in pepper production is the absence of an ideal ideotype that combines many positive traits to boost production potential while, at the same time, resisting the ravages of nature—such as the dreaded disease foot rot, caused by the *Phytophthora* fungi. The book also includes an exclusive chapter on a revolutionary soil management concept, developed by the author over three decades of research in Europe, Africa, and Asia, now globally known as the "nutrient buffer power concept," which has challenged the conventional wisdom of industrial agriculture, euphemistically known as the "green revolution." The book specifically details the work of the author on pepper nutrition with zinc, the most important micronutrient, which is suspected to be an intermediary in causing the *Phytophthora* onslaught in the quick wilt disease.

Malaparamba, Kozhikode, Kerala, India Kodoth Prabhakaran Nair

Acknowledgment

I wish to place on record a word of appreciation for Ms. Margaret Deignan, Senior Editor and Publisher, Springer, whose very thoughtful decisions and support led to the publication of this book. Ms. Neelofar Yasmeen, Production Editor (Books), provided diligent and very competent book editing. Ms. T. Metilda Nancy Marie Rayan, Project Manager, and her colleagues ably supported the printing of the book. I owe them all a deep debt of gratitude for doing a remarkable job.

Contents

Chapter 1
Introduction

Abstract The chapter traces the history of black pepper starting from the time of Queen of Sheba and King Solomon (BC 1015–BC 66) up to recent times. It lays stress on the fact that what the East India Company started as a trade in spices in India finally led to the colonization of the country. The chapter provides details of the spread of black pepper in the world, the acreage, the production, and the productivity. Among the Indian States, Kerala tops in black pepper production.

Keywords "King" of spices · British East India Company · Global distribution

Known as the "King" of spices, black pepper (*Piper nigrum*), a perennial crop of the tropics, is economically the most important and the most widely used spice crop of the world. The history of spices is very much entwined with the history of mankind. But, within the family of spices, black pepper predominates. In ancient Egypt, when the mummified body of the Pharaoh was laid to rest in the Pyramids, it was black pepper (along with gold and silver) that was kept adjacent to the body, in the belief that even in the afterlife this very important spice would be of use. The ancient scriptures, Bible, Koran, and the Vedas mention the use of spices. According to the Bible, it was during the royal visit of Queen Sheba to King Solomon (BC 1015–BC 66) that a caravan load of spices, primarily pepper, was presented by the former to the latter. Nearly 3000 years before the birth of Christ, both Babylonians and Assyrians were trading in spices, primarily black pepper, with the people of the Malabar Coast in the state of Kerala on the Indian subcontinent. Also, the ancient Indian medical texts, such as *Ashtangahridaya* and *Samhitas*, mention the use of pepper in rare and unique medical formulations. That spices, and in particular, pepper, had such a lasting impact on the economic prosperity of places is revealed by the fact that cities like Alexandria, Genoa, and Venice can trace their economic prosperity back to the vigorous trade in spices (Rosengarten 1973). Parry (1969) observed that due to the increased demand and consumption of pepper in England and Europe, a guild known as "Pepperers"—the wealthiest of the merchants—was established in London. The high price of pepper made it the exclusive commodity in

K. P. Nair, *The Geography of Black Pepper (*Piper nigrum*)*,
https://doi.org/10.1007/978-3-030-52865-2_1

use by the rich for culinary purposes. But, it is interesting to note that with the arrival of pepper, the Western kitchen transformed when "dishes took on a fullness of flavor previously unknown, beverages glowed with a redolent tang, and life experienced a new sense of warmth and satisfaction" (Parry 1969). King Solomon of Israel and the Phoenician King, Hiram of Tyre, obtained their spices from the coast of Malabar (Rosengarten 1973). Though the Jews and Arabs were well into the spice trade by this time, it was the latter who had a domineering role, which was later usurped by the Romans. However, the Arabs were privy to the knowledge of the country of origin of pepper and even secretly kept the sea route knowledge to themselves. But, the onslaught of the Roman Empire changed all that during the first century A.D. when Rome captured Egypt. It was around A.D. 40, during the reign of Roman Emperor Claudius that the enterprising Greek merchant mariner Hippalus, who—after discovering the full power and velocity of wind movement of the Indian ocean, a secret guarded by the Arabs—let it become known that a round trip to Egypt via India could be completed in less than 1 year (Rosengarten 1973). A consequence of the great economic scale of this discovery was that when the sea route came to be known, the dependence on the overland trade routes declined substantially; several wealthy cities on the overland trade route went into economic penury (Ummer 1989). Hippalus and many others reached the coast of Malabar and returned to Rome with loads of pepper and other spices. The Roman dominance on the spice trade completely eclipsed the Arabs. Interestingly, among all the spices, it was pepper that the Romans fancied most, despite its "obnoxious pungency" as noted by Pliny the Elder in his Natural History compiled between A.D. 23 and A.D. 79. It is also interesting to note that just about the time Arabs were getting involved in the spice trade with the Malabar Coast, the Chinese also entered the fray. As early as the first century A.D., a royal messenger reached the Malabar Coast in search of spices, and the trade relationship between China and the Malabar Coast began to flourish as recorded by travelers, such as Sulaiman, who reached the Malabar Coast in A.D. 851 (Ummer 1989). It was the Chinese who played an important part in spreading pepper to Southeast Asia and Far East Asia, and it was during the voyages of Zheng He (1405–1433) that China imported as much pepper as the total quantity imported into Europe during the first half of the tenth century (T'ien 1981). The Chinese trade was an imperial monopoly. In the fifteenth century, soldiers in China were partly paid in pepper, as were government officials, as observed by T'ien (1981). The fabulous Chinese ships that carried pepper were greatly praised by the venerable explorer Marco Polo (Mahindru 1982).

The colonization of the Indian subcontinent had much to do with the pepper trade. It was the lure of the spice trade that led Vasco de Gama, the great Portuguese explorer, to discover the sea route to India, and he landed on the Malabar Coast at the Kappad beach near the presently named city of Kozhikode (Calicut) on 20 May 1498. Vasco de Gama and his men returned to Portugal immensely rich, but in the process also put an end to the Arab trade. The Portuguese King, Dom Manuel, was so greatly impressed with the pepper bounty that Gama brought back home that he sent a naval contingent with a fleet of 13 battle ships to India under the command of Pedro Alvares Cabral in A.D. 1500; he went further to declare sovereignty over

India, along with other countries such as Ethiopia and Arabia (Rosengarten 1973). The Portuguese domination of Kerala through pepper production and trade was so complete that this tiny state of India turned out to be the cradle of world pepper. The Portuguese rulers were ruthless, and their only aim was to make the most out of the pepper trade. However, the scene changed in the first quarter of the seventeenth century with the arrival of the Dutch. The Dutch were temporarily successful in elbowing out the Portuguese until the British came on the scene in A.D. 1600. In hindsight, what is most astonishing is how a trade war in pepper between the Dutch and the British led to the establishment of the British Empire on the Indian subcontinent.

The historians Collins and Lapierre, in their monumental work *Freedom at Midnight* (1976), make a remarkable observation as follows.

Sometimes history's most grandiose accomplishments have the most banal of origins. Great Britain was set on the road to the great colonial adventure for five miserable shillings. They represented the increase in price of a pound of pepper proclaimed by the Dutch privateers who then controlled the spice trade. Incensed at what they considered a wholly unwarranted gesture, 24 merchants of the city of London gathered on the afternoon of 24 September 1599 A.D. in a decrepit building on Leadenhall Street. Their purpose was to find a modest trading firm with an initial capital of £72,000 subscribed by 125 shareholders. Only the simplest of concerns, that is, profit, inspired their enterprise, which expanded and transformed, would ultimately become the most noteworthy creation of the age of imperialism— the British Raj.

The British East India Company came into being on 31 December 1600 (with the stamp of approval by Queen Elizabeth I). Just 36 years later (1636), when the 500-ton ship *Hector* landed in the Surat port, north of Bombay, the Company laid the long and tortuous road to the subjugation of the vast millions of Indians through the pepper trade. In more than one sense, pepper was the cause of India losing its sovereignty.

It was only in the last quarter of the eighteenth century that the Americans entered the pepper trade. The first sponsored trip to the East Indies was organized by Capt. Jonathan Carnes in 1795. Though the American pepper trade flourished until 1810, it later declined, coinciding with the American Civil War in 1861. Compared with the Portuguese, Dutch, and British, the impact of Americans on the pepper trade was only marginal. The Americans traveled to Sumatra to fetch pepper. By 1933, pepper was introduced to Brazil, and in 1938, it reached the Republic of Malagasy. By 1954, pepper was introduced to the African continent.

Within the Indian Republic, the tiny state of Kerala can pride itself as the home of pepper, in particular, the coastal region of Malabar, in the state of Kerala, which accounts for 95% of the country's area and production (Anon. 1997). Besides Kerala, two other states in the southern region—namely, Tamil Nadu and Karnataka—contribute the remainder. The first research station of pepper in the world was established in Kerala in a small town named Panniyur on the Malabar Coast during 1952 and 1953. In addition, the first hybrid pepper—Panniyur-1—in the world was released by this station in 1966 (Fig. 1.1). Following research at

Fig. 1.1 Panniyur-1, the first pepper hybrid developed in the world at the Panniyur Pepper Research Station, in Kerala State, India

Panniyur, pepper research began in Sarawak (Malaysia) in 1955. Following the success of the "green revolution" in India, several coordinated research projects catering to the specific needs of individual crops were established by the Indian Council of Agricultural Research, headquartered in New Delhi. Thus, the All India Coordinated Research Project on Spices, with an intense focus on pepper, was established in 1971. More than 30 years later, the project has included other spice crops as well, yet pepper continues to receive the most attention against the background of the current liberalization and globalization process, and the World Trade Organization (WTO) mandated changes. The future of pepper is most crucial to the economy of India vis-à-vis the economy of the state of Kerala, where it is the economic mainstay.

Many other tropical countries have made concerted efforts to grow pepper in view of its global economic importance, the most important being Indonesia, where the International Pepper Community (IPC) is headquartered in Jakarta. Indonesia is second to India as a pepper-growing country. It was either through the Polynesian seafarers or through the Babylonian-Chinese sea route linking the Malabar Coast and southeast and Far East Asia that pepper reached Indonesia. Indonesia, known as the Dutch East Indies during the pre-World War II period, was the largest pepper producer. It was during the Japanese occupation during the war that many plantations were abandoned and production declined sharply. During the prewar period, Indonesia had close to 30,000 ha of pepper (Lawrence 1981). Malaysia and Sri Lanka are the other two major pepper producers. It was the European settlers who introduced pepper to Malaysia, while in Sri Lanka, pepper was grown as a mixed crop with other crops like cocoa. The foreign occupation of Sri Lanka helped pepper cultivation expand in the country. In the Southeast Asian countries such as Thailand, Vietnam, Cambodia, South Korea, and parts of South China, pepper cultivation took

Table 1.1 Acreage, production, and productivity of major pepper-growing countries in the last decade of the twentieth century

Country	Area (ha)	Production (metric tons)	Productivity (kg ha^{-1})
Brazil	26,500	23,400	883
China	13,170	11,045	839
India	191,426	56,200	294
Indonesia	110,580	45,240	409
Madagascar	4228	2160	511
Malaysia	8960	16,920	1888
Mexico	1294	1112	859
Sri Lanka	12,080	5058	419
Thailand	2808	10,091	3594
Vietnam	15,700	17,266	1100

hold in the postwar years. Within the Southeast Asian countries, Vietnam is beginning to emerge as a major pepper grower. Within the South Pacific islands, Fiji is the most important pepper-growing country. In South America, Brazil is the leader, followed by Mexico, Guatemala, Honduras, Saint Lucia, Costa Rica, and Puerto Rico. Within the African continent, Madagascar leads the pack followed by Malawi, Zimbabwe, Benin, Kenya, Nigeria, Cameroon, Congo, and Ethiopia.

Within Asia, where pepper production is concentrated, Malaysia takes third place next to Indonesia, with an Agricultural Research Station in Kutching, Sarawak. Indonesia, which takes second place, has the Research Institute for Spice and Medicinal Plants in Bogor. Though India tops the list among the producers in acreage, with a total area of 191,426 ha and a production of 56,200 metric tons, its productivity is the lowest in the world with just 294 kg ha^{-1}, while Thailand, with a total area of only 2808 ha, tops the list with a productivity of 3594 kg ha^{-1} (Table 1.1). The International Trade Center (ITC) in Geneva, Switzerland, estimates the current trade in spices at 400,000–450,000 metric tons with a total value of US $1.5–$2 billion annually. With an annual growth rate of 3.6% in quantity and 8.4% in value in spices, pepper contributes 34% of total trade in spices. Within the industrialized west, Denmark tops the list in pepper consumption, followed closely by Germany and Belgium. The USA is a sizable consumer, while Canada and Switzerland tail the list.

Post-Doha (World Trade Organization (WTO) held an important meeting in Doha, Qatar, on global trade regulations, including agriculture) negotiations, agriculture will increasingly play a crucial role in the world economy. Among the spices, pepper will play a major role. By 2010, projected world consumption will reach 230,000 metric tons, which scales up to 280,000 metric tons by 2020, which means an annual increase of 5000 metric tons. Present production is close to 200,000 metric tons. For the next two decades, close to 100,000 metric tons will be needed to balance the projected demand and consumption. Worldwide, especially in the industrialized countries, there is a growing demand for premium organically grown pepper. The potential for the organic food market is close to US $8 billion now in the USA, which is followed by Germany and Japan, each with a market share of close to US $2.5 billion. A substantial part of this market will be for pepper. This would imply that future production strategies would need to increasingly focus on clean pepper production, which has to withstand both biotic and abiotic stresses without recourse to high input chemical technology—the hallmark of the so-called green revolution.

There are areas of pepper production that simultaneously pose great challenges while opening up new avenues. One of the most daunting in the former category is the evolution of a pepper variety that is totally resistant to "foot rot," caused by the fungus *Phytophthora*, which has wiped out many pepper plantations. Also, pepper nutrition is still far from being thoroughly understood. The fact that it is a perennial crop adds to the lack of thorough understanding. Despite the complexity of soil science and emergent soil management practices, the basic concept of soil as a medium of plant growth can be expected to persist for an indefinite length of time (Nair 1996). But it is becoming clearer that the earlier views on soil as merely a "supportive medium" for plant growth is giving place to new ones on managerial concepts of this supportive medium. This is amply illustrated by the shift in focus from the green revolution phase of the 1960s to mid-1970s—where application of increasing quantities of soil inputs, such as fertilizers and pesticides, was emphasized—to the sustainable agriculture phase of the early 1980s continuing to the present; sustainable agriculture places more reliance on biological processes by adopting genotypes to adverse soil conditions, enhancing soil biological activity, and optimizing nutrient cycling to minimize external inputs (such as fertilizers) and maximize their efficiency of use. In fact, the paradigm of the earlier phase has given way to an emergent new paradigm (Sanchez 1994), and this is clearly reflected in the dialogue of the world leaders during the Earth Summit in 1991 in Rio de Janeiro, Brazil, where Agenda 21 incorporated six chapters on soil management (Keating 1993). This review on pepper, while discussing its overall production profile in the world, will place a special emphasis on the second paradigm inasmuch as prescriptive soil management for pepper production is concerned with regard to understanding the soil nutrient bioavailability and its efficient management in the pepper production. On account of the paucity of published literature on this aspect, the focus will only be with regard to specific nutrients, such as zinc, which is becoming increasingly important in pepper production.

References

Anon (1997) Spice export grows to an all time record. Spice India 10:2–4

Collins L, Lapieree D (1976) Freedom at midnight. Vikas Publication House, New Delhi

Keating M (1993) The Earth Summit's agenda for change. In: A plain language version of Agenda 21 and the other Rio Agreements. Centre for our common future, Geneva

Lawrence BM (1981) Major tropical spices—pepper (*Piper nigrum* L.). In: Essential oils 1979–80. Allured Publications Corporation, Wheaton, pp 141–228

Mahindru SN (1982) Spices in Indian life. Sultan Chand & Sons, New Delhi

Nair KPP (1996) The buffering power of plant nutrients and effects on availability. Adv Agron 57:237–287

Parry JW (1969) Spices, vol 1. Chemical Publishing Co. Inc., New York

Rosengarten F Jr (1973) The book of spices. Pyramid Books, New York

Sanchez PA (1994) Tropical soil fertility research: towards the second paradigm. In: Proceedings of the XV ISSUS Symposium, Acapulco, Mexico, July 10–16, vol 1, pp 89–104

Tien J-K (1981) China and the pepper trade. Hemisphere 18:220–222

Ummer C (1989) Indian spices-from the leaves of history. Spice fair commemorative volume. Spices Board, Cochin, pp 27–40

Chapter 2
The Pepper Plant: Its Botany and Chemistry

Abstract The chapter details the botany and chemistry of the pepper plant. Details of the black pepper varieties, their origin, and average yield are catalogued. The section on black pepper chemistry focuses on the distinctive pungency and aroma of pepper. This has both intrigued and fascinated pepper chemists. Different researchers have reported wide variation in the chemical composition of essential oils in black pepper, and these variations originate because of several reasons, such as varietal differences, their geographic origin, variations in the maturity of raw material, procedural differences in extraction, and nonresolution of constituents in early gas chromatographic analysis employing packed columns.

Keywords Botanical names · Pepper varieties · Pepper chemistry

2.1 Pepper Botany

Black pepper (*P. nigrum*) is a perennial plant and derives its name—"Piper"—perhaps from the Greek name for black pepper, *Piperi* (Rosengarten 1973); most of the European names for black pepper were derived from the ancient Indian language Sanskrit, such as *Pippali*, the name for long pepper (*P. longum*). It was the great botanist Linnaeus (1753) who established the genus *Piper* in his work *Species Plantarum*. In this monumental work, Linnaeus recognized 17 species in the Piper family, all of which were included in the same genus. In 1974, Ruiz and Pavon introduced the second genus in the family, namely, Peperomia (Trelease and Yuncker 1950). The family name Piperaceae was used for the first time in 1815 by L. C. Rich in Humboldt, Bonpland, and Kunth's *Nova Genera et Species Plantarum* (Yuncker 1958). All the species known in the family Piperaceae during the early years of systematic classification were included in the classical monographic study *Systema Piperacearum* published in 1843 by F. A. W. Miquel.

It was Rheede who made the earliest record of the description of Piper in the Indian subcontinent in 1678. In his *Hortus Indicus Malabaricus*, the earliest printed document of plants on the Malabar coast, Rheede described five types of wild pepper

K. P. Nair, *The Geography of Black Pepper (*Piper nigrum*)*,
https://doi.org/10.1007/978-3-030-52865-2_2

that included both black and long pepper. Linnaeus (1753) included 17 species from India in his monumental work *Species Plantarum*. The first major study of the *Piper* spp. from the Indian subcontinent was that of Hooker (1886) in his book *Flora of British India*. However, the most authoritative floristic study of the Western Ghats of southern India was that of Gamble in 1925 in his book *Flora of Presidency of Madras*, in which the following 13 species with their taxonomic keys were described: *P. argyrophyllum*, *P. attenuatum*, *P. barberi*, *P. brachystachyum*, *P. galeatum*, *P. hapnium*, *P. hookeri*, *P. hymenophyllum*, *P. longum*, *P. nigrum*, *P. schmidtii*, *P. trichostachyon*, and *P. wightii*. Following the publication of Gamble's text, no new additions to the list were made until 1981, when a new species, *P. baba-budani*—from the Baba Budan hills of the state of Karnataka (the neighboring state of Kerala)—was added (Rahiman 1981); however, this was not published. Ravindran et al. (1987) reported a new species, *P. silentvalleyensis*—the only bisexual wild species from the world-renowned Silent Valley in the Western Ghats. Other new species are *P. pseudonigrum* (Velayudhan and Amalraj 1992) and *P. sugandhi* (Nirmal Babu et al. 1993a).

Black pepper cultivars could possibly have originated from wild ones through the process of domestication and selection. More than a hundred cultivars are known, but most have vanished due to the onset of devastating diseases like foot rot and replacement by hybrids. Human migration has contributed to the spread of these cultivars. Pepper distribution is extensive in moist evergreen forests and, to a lesser extent, in semievergreen and moist deciduous forests of the Western Ghats of southern India, growing from almost sea level to a height of 1500 m above mean sea level. Population structure of any species is determined mainly by the breeding system of the species; by the pollen, fruit, and seed dispersal mechanism; and the presence or absence of isolation mechanisms. In *Piper*, male, female, and hermaphrodite forms exist. The cultivated *P. nigrum* is monoecious, having hermaphrodite flowers, while the wild ones are mostly dioecious. Human selection has played a major role in the directional evolution of hermaphroditism in the cultivated pepper. Predominantly self-pollinated, pepper pollen dispersal is aided by rain or dewdrops and also by geitonogamy—the gravitational descent of pollen. Though flowers are protogynous, in the absence of an active pollen transfer mechanism, protogyny is ineffective in outbreeding. Active and efficient pollen and seed dispersal mechanisms ensure gene flow within and between population segments, leading to the establishment of intergrading populations. Absence of such a mechanism in *Piper* ensures effective isolation barriers among individuals and population units. Segregation in the seedling progenies leads to variations in such units, as well as mutations and chance crossing followed by segregation. Any such variation stemming in population is fixed immediately because of the vegetative mode of propagation, and such a unit may, over the course of time, diverge from other similar units. Quite often, different types of *P. nigrum* or different species climb up a single tree, which enhances the chances of outcrossing, which results in hybrid seedlings. Progenies from such chance crosses grow and later climb up the same or nearby trees and chance outcrossing with the parental vine or its clonal or other seedling progenies, which result in further back crossing or hybrid progenies—all of which lead to substantial variation within the population. These forces acting together would have most likely contributed to the evolution of many present-day pepper cultivars (Ravindran et al. 1990).

Though pepper has originated in the evergreen forests of the Western Ghats in southern India, with extensive occurrence of wild pepper in the less disturbed forest areas, no study has so far been carried out to find out the origin of *P. nigrum*. The basic chromosome number of *Piper* is $x = 13$ and that of *P. nigrum* $2n = 52$, a tetraploid (Mathew 1958, 1972; Jose and Sharma 1984). On the basis of morphological and biosystematic studies, Ravindran (1991) reported that three species—namely, *P. wightii*, *P. galeatum*, and *P. trichostachyon*—could be the putative parents of *P. nigrum*. These three are woody climbers with similar texture and leaf morphology. Their spikes and fruits are more similar to that of *P. nigrum* than those of other species. Of the three, P. wightii and P. galeatum are the most probable ancestors of P. nigrum.

Pepper is a shade-loving plant that requires a constant moisture supply during dry spells because it has a high evapotranspiration coefficient (Raj 1978). However, even under favorable soil moisture conditions, when exposed to direct sunlight, the plant develops certain physiological disorders (Vijayakumar et al. 1984). Pepper is highly sensitive to the growth-light regime (GLR), and plant parts exposed to high GLR produce more fruits per unit surface area than those exposed to low GLR (Montaya et al. 1961). Where grown in permanently shaded situations, productivity is poor. Under such conditions, plants respond positively to radiation, and a positive correlation has been shown to exist between productivity and radiation. Except in the case of "bush pepper," normal pepper vines are grown on artificial support. These supports are called "standards." Ramadasan (1987) reports three types of pepper canopies. Canopy shaping is largely based on the shading provided by live support and that of the adjoining shading trees. When dead supports such as brick pillars or reinforced concrete poles are used, as in Thailand, the canopy does not taper at the top. Trees that grow adjacent to dead supports lead to tapering of the canopy at the base due to partial shading. Such tapering at the base is not observed when there are no competing trees. In Sarawak, the wood of Bornean ironwood (*Eusideroxylon zwageri*) is used for support, and generally, no shade tree is grown. In such dead supports, the top of the canopy unfolds like an umbrella.

The pepper plant is endowed with two advantages that many others do not have. First, it can be vegetatively propagated; second, it is viable for sexual reproduction. Both offer a large scope for the exploitation of hybrid vigor as well as selection breeding. Published literature show that clonal selection, hybridization, open pollinated progeny selection, mutation, and polyploidy have been successfully experimented with to improve pepper yield and impart desirable plant traits. In recent times, biotechnological methods are also being employed to impart pathogen resistance.

Quality is the most important criterion in breeding as far as spices are concerned, and pepper is no exception. The quality of pepper is decided mainly by the amounts of piperine and oleoresin (essential oil). An alkaloid, piperine contributes to the pungency of the berry, and oleoresin content enhances the flavor. The unique pepper flavor is influenced by the large number of chemical compounds present in the essential oil, and a number of genes control this unique character.

Conservation of pepper genetic resources is the key to sound pepper breeding. Among the 18 biodiversity hot spots on this planet, the Western Ghats of southern India is one of the most unique, with its different forest types—such as tropical wet evergreen, tropical moist and dry deciduous, and montane subtropical and temperate ones. It is on the western side of the Western Ghats that the population and diversity of *Piper* is in abundance. It is important to realize that during the past century, the Western Ghats have been ravaged the most (leading to ecological breakdown) due to clandestine timber felling and uncontrolled human encroachment. The Western Ghats is home to spices that are endemic to the region. Of the 17 species of *Piper* found here, 11 are endemic to the region (Ravindran and Peter 1995). More than a hundred cultivars were under cultivation in the region, but most of them have vanished from the pepper growers' gardens, probably due to foot rot (*Phytophthora*) and slow decline as well as replacement of old and unproductive cultivars by the few high-yielding elite ones. It takes millennia for nature to nurture plant diversity, but human activity, productive or otherwise, can change all that in a very short span of time.

As part of the conservation efforts, some pepper-growing countries, notably India, have made a concerted effort to collect and catalogue pepper germplasm in a systematic manner. Mention should be made of the efforts in this direction undertaken at the Indian Institute of Spices Research at Kozhikode, Kerala State. The Institute has established a National Conservatory for in situ conservation. Conservation is carried out in four stages (Ravindran and Nirmal Babu 1994) as follows: (1) conservation as a nursery gene bank by trailing each accession in split-bamboo pieces of about 2–3 ft long serially, which will be maintained and continuously multiplied; (2) conservation in the clonal repository, where ten rooted cuttings from each accession are maintained; (3) conservation in the field, as a field gene bank, by planting accessions for pilot yield performance and characterization; and (4) conservation in vitro and in cryogen banks. With the establishment of the All India Coordinated Research Project on Spices in 1971, collection and conservation of pepper germplasm on a smaller scale has also been made by the Panniyur Pepper Research Station (the first pepper research station established in the world) in the state of Kerala; the Pepper Research Station at Sirsi, in the state of Karnataka; the Horticultural Research Station, Chintapalli, in the state of Andhra Pradesh; and the Horticultural Research Station, Yercaud, in the state of Tamil Nadu. Among these centers, the research station in Yercaud has the largest collection, but it is the Indian Institute of Spices Research that can pride itself as the one holding the largest collection in the world. On a much smaller scale, collections and conservation have also been made, notably in Indonesia and Malaysia, with the former holding 40 cultivars and 7 *Piper* species and Sarawak (Malaysia) holding 18 cultivars of *P. nigrum*, 18 identified *Piper* spp., and 98 unidentified accessions (De Waard 1984). Despite the fact that they are not closely related to the cultivated black pepper, the vast number of *Piper* spp. found in central and southern America, northeastern India, Indonesia, and Malaysia have not been collected and

Table 2.1 Some improved pepper varieties from major pepper-growing countries

Country of origin and name of variety	Average yield (kg ha^{-1})	Remarks
India		
Panniyur-1	1242	Well suited to most regions; performs rather poorly at higher elevations and under shade
Panniyur-2	2750	Well suited to most regions; shade-tolerant
Panniyur-3	1953	Well suited to most regions; rather late in maturity
Panniyur-4	1277	Consistently good yielder; suited to most regions
Panniyur-5	1098	Most important characteristic is tolerance to nursery diseases and shade
Subhakara	2352	Suited to most regions; berries of good quality
Sreekara	2677	Suited to most regions; berries of good quality
Panchami	2828	Suited to most regions; late in maturity
Pournami	2333	Good tolerance to root knot nematode
PLD-2	2475	Suited to most regions; berries of high quality
Indonesia		
Natar 1	a	High yield; particularly tolerant to foot rot and nematodes
Natar 2	–	High yield; particularly tolerant to foot rot and nematodes
Malaysia		
Semongok Perak	–	High yield
Semongok Emas	–	Uniquely high yield; tolerant to foot rot, black berry disease, and pepper weevil
Madagascar		
Sel. IV.1	–	High yield
Sel. IV.2	–	High yield
Sri Lanka		
PW 14	–	Claimed to be resistant to *Radopholus similis*

[a]Average yield unknown

systematically catalogued. The vast germplasms in these regions could form a very important reservoir of gene pool to develop varieties with inbuilt disease and pest resistance (Table 2.1).

A new avenue for pepper improvement is through biotechnological research, which of late has made great strides in many areas of biology. The biotechnological approaches permit the researchers to manipulate plant tissues and cells in vitro. A better understanding of genetics is provided by tissue culture techniques and investigations at the molecular, cellular, and organismal levels. The possibilities of cell culture and plant tissue culture for breeding and plant propagation purposes have led to large-scale commercial exploitation. Technical perfection by manipulating plant cells at the molecular level through recombinant DNA technology has opened up enormous possibilities in the creation of transgenic plants, and this is an area that is assuming industrial proportions. Inasmuch as pepper is concerned, one area where recombinant DNA technology could be successfully exploited is the evolution of pepper varieties tolerant to foot rot caused by *Phytophthora capsici*. None of the varieties in cultivation is tolerant to this disease. Within the *Piper* spp., a distant relative of *P. nigrum*—namely, *P. colubrinum* from South America—is resistant to *Phytophthora*, and, hence, biotechnological approaches that will enable transfer of resistance-governing genes from *P. colubrinum* to *P. nigrum* will be the only mode to contain the devastation from this dreaded disease.

Another area of interest is micropropagation, where in vitro culture methods for pepper cloning is done. Micropropagation was first introduced by Broome and Zimmerman (1978). For micropropagation, several plant parts—such as shoot tips, nodal segments, and apical meristems—from both juvenile and mature plants have been used (Agarwal 1988; Joseph et al. 1996; Lissamma et al. 1996; Mathews and Rao 1984; Nazeem et al. 1993; Nirmal Babu et al. 1993a, b; Philip et al. 1992). However, a serious limitation to micropropagation is the incidence of bacterial contamination (Kelker and Krishnamoorthy 1996), though contamination is minimal in seedling explants. The authors (Kelker and Krishnamoorthy 1996) suggest the incorporation of antibiotics in the culture media to arrest the endogenous bacterial contamination in the in vitro pepper cultures. Field establishment of micropropagated pepper is feasible, though endogenous bacterial contamination and phenolic exudates from the cut surface could adversely affect field establishment (Fitchet 1988a, b; Raj Mohan 1985). The authors suggest treating the explants with fungicides prior to routine sterilization followed by frequent transfer to fresh medium to keep in check the problem of contamination. Madhusudhanan and Rahiman (1996) suggest the use of activated charcoal, at the rate of 200 mg l^{-1}, which could reduce browning of the explants and culture medium. Micropropagation has been in use for quite some time (Chua 1981; Fitchet 1988a, b). Culture media commonly used in micropropagation are those of Murashige and Skoog (1962), Schenk and Hildebrandt (1972), and McCown and Amos (1979). Indole acetic acid (IAA), naphthalene acetic acid (NAA), and 2,4-dichlorophenoxy acetic acid (2,4-D) are some of the commonly used growth regulators and additives used in the culture formulation.

One other area of pepper improvement is the *Agrobacterium*-mediated gene transfer. Reference to the technique is made by Sasikumar and Veluthambi (1994) and Sim et al. (1995). The technique can also be helpful in transferring disease resistance genes from *P. colubrinum* to *P. nigrum* in the light of the foot rot disease. Work on transgenic pepper to evolve delayed ripening ensuring uniform maturity—which will ensure reduction in labor costs for harvest—in Sarawak (Malaysia) has led to promising results.

Conservation of pepper germplasm assumes high practical importance in improving the crop for enhanced productivity and acquisition of many desirable traits, the most important being tolerance to diseases such as *Phytophthora*-caused foot rot. Normal conservation of crop germplasm is through seed banks. Because pepper is propagated only vegetatively (seeds are heterozygous and unpredictable in behavior), in vitro storage of germplasm is a viable alternative. In vitro conservation of pepper germplasm and its related species—such as *P. barberi*, *P. colubrinum*,

P. beetle, and *P. longum*—by defining protocols that have been made by Geetha et al. (1995) and Nirmal Babu et al. (1996) by maintaining cultures at reduced temperatures in the presence of osmotic inhibitors at low nutrient levels and also by reducing evaporative losses to a minimum using closed containers. When the conserved materials were transferred to a multiplication medium after storage, all species exhibited normal multiplication rates. When the normal-sized plantlets were transplanted into field soil, an establishment rate exceeding 80% was obtained. Normal pepper plants, morphologically similar to mother plants, were observed to develop.

There is a steep decline in viability in pepper seeds associated with moisture depletion. Seeds cryopreserved in liquid nitrogen at – 196 °C with an initial moisture percentage of 12 have a survival rate of only 45%, while at 6% moisture, it reduced to just 10.5% (Chaudhary and Chandel 1994), which clearly shows how seed moisture is crucial to seed survival. There is yet another manner in which pepper plants can be propagated: by producing synthetic seeds, which consist of somatic embryos or shoot buds encapsulated in a protective coating that is biodegradable. The system facilitates not only low-cost propagation but also germplasm conservation and exchange. Encapsulation of disease-free bud employing tissue culture techniques helps production of disease-free plantlets. Pepper shoot buds 0.5 cm long were used for production of synthetic seeds, which could be stored up to 9 months in sterile water. The method is described by Sajina et al. (1996).

2.2 Pepper Chemistry

It is the unique pungency and aroma of pepper that has both intrigued and fascinated pepper chemists. As already mentioned, the essential oil in the berry—oleoresin—contributes to the aroma, while the alkaloid piperine imparts the unique pungency. Oleoresin has very great commercial value and is extracted from the dry powdered berries. Dramatically put, it is the chemistry of pepper emanating from oleoresin that imparts that rare and unique blend of both aroma and flavor that has been the cause of wars and enslavement (as much as love and tragedy), which has been detailed in the introductory part of this review. The earliest investigations on pepper chemistry go back more than half a century (Guenther 1950), and subsequent research has enriched it (Govindarajan 1977; Lawrence 1981; Purseglove et al. 1981; Wealth of India 1969).

The various compounds occurring in *Piper* spp. are listed by Parmar et al. (1977). It was in the early nineteenth century that the first report on essential oil in pepper was made by Dumas and later by Subeiran and Capitaine (Guenther 1950). These researchers concluded that pepper oil is almost free of oxygenated constituents. By treating a fraction of the oil boiling at 176 °C with acid and alcohol, Eberhardt obtained terpin hydrate and the presence of 1-phellandrene, caryophyllene, and tentatively dipentene were reported by Schimmel and company and by Schreiner and Kremers (Guenther 1950). The steam distillation of essential oil obtained from dry powdered Malabar pepper berries showed the presence of α-pinene, β-pinene, 1-α-phellandrene, dl-limonene, piperonal, dihydrocarveol (a compound melting at 161 °C), β-caryophyllene, and a piperidine complex (Hasselstrom et al. 1957). Additionally, the presence of cryptone, epoxydihydrocaryophyllene, and, possibly, citronellol and an azulene were also reported. The presence of α- and β-pinenes, limonene, and caryophyllene in the hydrocarbon part of black pepper oil was confirmed by infrared spectroscopy (Jennings and Wrolstad 1961).

It is important to mention that a renewed thrust was given to the study of chemical compounds in pepper after the advent of gas chromatography. Modern researchers used thin-layer chromatography, gas chromatography, column chromatography, vacuum distillation, and the like to separate the constituents and employed ultraviolet, infrared, nuclear magnetic resonance, and mass spectroscopy for identification.

As many as 135 compounds—consisting of monoterpenoids, sesquiterpenoids, aliphatic, aromatic, and those of miscellaneous nature—have been reported (Debrauwere and Verzele 1975a, b, 1976; Gopalakrishnan et al. 1993; Ikeda et al. 1962; Lawrence 1981; Muller and Jennings 1967; Muller et al. 1968; Nigam and Handa 1964; Richard and Jennings 1971; Russel and Else 1973; Sharma et al. 1962; Wrolstad and Jennings 1965).

Different researchers have reported wide variations in the chemical composition of essential oils in pepper, and these variations originate for several reasons, such as varietal differences, their geographic origin, variations in the maturity of raw material, procedural differences in oil extraction, and nonresolution of constituents in early gas chromatographic analysis employing packed columns. To a certain extent, composition of essential oils will depend on the method of preparation; for example, steam distillation will produce oils containing about 70–80% monoterpene hydrocarbons, 20–30% sesquiterpene hydrocarbons, and less than 4% oxygenated constituents. Vacuum-distilled oils will contain less monoterpene hydrocarbons and more sesquiterpene hydrocarbons and oxygenated constituents. It is the result of incomplete distillation and the poor recovery of the high boiling sesquiterpene hydrocarbons and oxygenated constituents that lead to their low contents in steam-distilled oils compared with vacuum-distilled oils. Seventeen pepper cultivars from the state of Kerala were analyzed for their essential oil contents by Lewis et al. (1969) and Richard et al. (1971), and comparable results were obtained. Gopalakrishnan et al. (1993) analyzed the four Panniyur genotypes (Panniyur-1, Panniyur-2, Panniyur-3, and Panniyur-4) developed at the Pepper Research Station in Panniyur using a combination of gas chromatography and mass spectroscopy and Kovats indices on a methyl silicone capillary column and found, on the whole, that Panniyur-4 contained the most chemical constituents as compared with the other three genotypes. Significant differences could also be found in 12 samples from Lampong and 16 samples from Sarawak, both in Malaysia, using the procedure of Richard et al. (1971) by Russel and Else (1973).

An interesting aspect of research refers to sensory evaluation of essential oils in pepper. There are but scanty references with regard to the relationship between odor characteristics and oil composition. The characteristic odor of pepper oil has been attributed to the meager amounts of oxygenated constituents (Hasselstrom et al.

1957). Arctander (Purseglove et al. 1981) describes the odor of pepper oil as fresh, dry-woody, warm-spicy, and identical to that of black pepper corn. Pangburn et al. (1970), after a systematic and thorough sensory evaluation of pepper oil obtained from pepper along the Malabar coast and using column chromatographic fractions and mixtures of fractions of the oil, reported that the early fraction was considered pepper-like and floral; the late fraction pepper-like, fresh, and woody; and the middle fraction remaining in between. Direct sniffing at the eluting port of the gas chromatographic columns helped distinguish the distinct odor of black pepper and ascribe it to the three areas of the late fractions. Distinctive odor analysis has been developed by Harper et al. and Sydow et al. (Govindarajan 1977). Using a descriptive odor profile based on a four-point category scale and subjecting the oils to a ranking list, Gopalakrishnan et al. (1993) have described the odor evaluation of the four Panniyur genotypes described earlier in this review.

Besides aroma (odor), the other aspect that has been of interest to the researchers is the pungency emanating from the alkaloid piperine. From the early nineteenth century onward, the pungency of pepper has been investigated—starting with the first report of Oersted, who isolated piperine in 1819 (Guenther 1950). Piperine is a yellow crystalline substance that was subsequently identified as the *trans* form of piperoyl piperidine. Subsequent researchers showed that piperine was not the only substance imparting pungency to pepper, and Buccheim (Govindarajan 1977) obtained a dark oily resin, which he called "chavicine," subsequent to the removal of piperine from the oleoresin. Chavicine was supposedly far more pungent than piperine, leaving a much sharper bite on the tongue than the crystalline piperine. This was shown to be incorrect by subsequent researchers who demonstrated the extreme pungency of piperine in solution. The debate, whether it was chavicine or other possible isomers, such as isopiperine or isochavicine, which are more pungent than piperine, continued for almost a century. However, recent investigations reveal that it is piperine that is the most pungent; chavicine is a mixture of piperine and many other minor alkaloids.

The question of why pepper blackens has interested many researchers. Lewis et al. (1976) attributed blackening of pepper berries to enzymatic oxidation of polyphenolic substrates present in the skin of green pepper. The blackening of pepper is a chemical reaction akin to browning in fruits and vegetables and can have an enzymatic or nonenzymatic origin. Formation of coloring pigments because of enzymatic browning is triggered by the enzyme polyphenol oxidase. The authors Lewis et al. (1976) developed a process in which green pepper was blanched to arrest the enzyme reaction with subsequent drying in a cross-flow dryer to obtain dehydrated green pepper. Pruthi et al. (1976) reported that green pepper could also be preserved in brine containing either acetic or citric acid.

In addition to the previously described chemical constituents, starch is a predominant constituent of pepper; its content may vary between 35% and 40% in terms of weight. In addition to starch, pepper also contains protein and fat. Reports on both starch and protein are practically nonexistent. As for fat, pepper contains 1.9–9% fat. Salzer (1975) has determined the fatty acid composition of pepper and reported that of the different fatty acids, such as palmitic, oleic, linoleic, and linolenic, it is the linoleic acid that is most abundant (25–35%) and linolenic is the least abundant (8–19%).

The survey of pepper chemistry shows that though some of the major constituents have been researched during the last few decades, in the future, one may still come across constituents hitherto unknown. The research so far has mostly been confined to the varieties in cultivation in different regions worldwide, and yet there are far more cultivars that remain totally out of the chemists' laboratory. It is important to research these, because one may come across unique chemical compositions and flavor profiles in the different cultivars. As with disease resistance, unique chemical constitution is also of immense commercial importance. Only when pepper breeders, biotechnologists, and pepper chemists join hands in an intense search for unique traits will the research on pepper chemistry have any meaningful outcome, because it is only through these intense efforts that unique pepper lines could be identified and isolated for future breeding and crop improvement programs.

References

Agarwal S (1988) Shoot tip culture of pepper and its micropropagation. Curr Sci 57:1347–1348

Broome OC, Zimmerman RN (1978) *In vitro* propagation of black pepper. Hortic Sci 43:151–153

Chaudhary R, Chandel KPS (1994) Germination studies and cryo preservation of seeds of black pepper (*Piper nigrum* L.) a recalcitrant species. Cryo-Lett 15:145–150

Chua BK (1981) Studies on *in vitro* propagation of black pepper (*Piper nigrum* L.). MARDI Res Bull 8:155–162

De Waard PWF (1984) *Piper nigrum* L. In: West Phal E, Jansen PCM (eds) Plant resources of South East Asia. Rijksherbarium, Leiden, pp 225–230

Debrauwere J, Verzele M (1975a) Constituents of pepper VI. Oxygenated fraction of pepper essential oil. Bull Soc Chem Belg 84:167–177

Debrauwere J, Verzele M (1975b) New constituents of the oxygenated fraction of pepper essential oil. J Sci Food Agric 26:1887–1894

Debrauwere J, Verzele M (1976) The hydrocarbons of pepper essential oil. J Chromatogr Sci 14:296–298

Fitchet M (1988a) Establishment of black pepper in tissue culture. In: Information Bulletin No. 189. Citrus and Subtropical Fruits Research Institute, Nelspruit

Fitchet M (1988b) Progress with *in vitro* experiments of black pepper. In: Information Bulletin No. 196. Citrus and Subtropical Fruits Research Institute, Nelspruit

Geetha SP, Manjula C, Sajina A (1995) *In vitro* conservation of genetic resources of spices. In: Proceedings of the seventh Kerala Science Congress, January 27–29. Science Technology and Environment Committee, Kerala Government, pp 12–16

Gopalakrishnan M, Menon N, Padmakumari KP, Jayalakshmi A, Narayanan CS (1993) GC analysis and odor profiles of four new Indian genotypes of *Piper nigrum* L. J Essent Oil Res 5:247–253

Govindarajan VS (1977) Pepper—chemistry, technology and quality evaluation. Crit Rev Food Sci Nutr 9:115–225

Guenther E (1950) Essential oils of the plant family Piperaceae. In: The essential oils, vol 5. Robert Krieger Pub. Co., New York, pp 135–161

Hasselstrom TF, Hewitt EJ, Konigsbacher KS, Ritter JJ (1957) Composition of oil of black pepper. J Agric Food Chem 5:53–55

Hooker JD (1886) The flora of British India, vol 5. L. Reeva and Co., London, pp 79–95

Ikeda RN, Stanley WL, Vannier SH, Spitler FM (1962) The monoterpene composition of some essential oils. J Food Sci 27:455–458

Jennings WG, Wrolstad RE (1961) Volatile constituents of black pepper. J Food Sci 26:499

Jose J, Sharma AK (1984) Chromosome studies in the genus Piper L. J Indian Bot Soc 63:313–319

Joseph B, Joseph D, Philip VJ (1996) Plant regeneration from somatic embryos in black pepper. Plant Cell Tissue Org Cult 47:87–90

Kelker SM, Krishnamoorthy KV (1996) Control of bacterial contamination in *in vitro* cultures of pepper (*Piper* sp.). In: Proceedings of the national seminar on biotechnology of spices and aromatic crops, April 24–25, 1996, Calicut, Kerala State, India

Lawrence BM (1981) Major tropical spices—pepper (*Piper nigrum* L.). In: Essential oils 1979–80. Allured Publications Corporation, Wheaton, pp 141–228

Lewis YS, Nambudiri ES, Krishnamurthy N (1969) Composition of pepper oil. Perfum Essent Oil Rec 60:259–262

Lewis YS, Krishnamurthy N, Nambudiri ES, Sankarikutty Amma B, Shivshankar S, Mathew AG (1976) The need for growing pepper cultivars to suit pepper products. Proceedings of the international seminar on pepper. Spices Export Promotion Council, Cochin, pp 4–9

Linnaeus C (1753) Species plantarum, London

Lissamma J, Nazeem PA, Mini ST, Shaji P, Mini B (1996) *In vitro* techniques for mass multiplication of black pepper (*Piper Nigrum* L.) and *ex vitro* performance of the plant- lets. J Plant Crops 24(Suppl):511–516

Madhusudhanan K, Rahiman BA (1996) *In vitro* response of Piper species on activated charcoal supplemented media. In: Edison S, Ramana KV, Sasikumar B, Nirmal Babu K, Eapen SJ (eds) Biotechnology of spices, medicinal and aromatic plants. Indian Society for Spices, Calicut, pp 16–19

Mathew PM (1958) Studies on Piperaceae. J Indian Bot Soc 37:155–171

Mathew, P. M., 1972. Karyomorphological studies on *Piper nigrum.* In: Proceedings of the national symposium on plantation crops. J Plant Crops 1 (Suppl), 15–18.

Mathews VH, Rao PS (1984) *In vitro* responses in black pepper (*Piper nigrum*). Curr Sci 53:183–186

McCown HB, Amos R (1979) Initial trials of micropropagation with birch. Proc Int Plant Prop Soc 29:387–393

Montaya LA, Sylvain PG, Umana R (1961) Effect of light intensity and nitrogen fertilization upon growth differentiation balance in *Coffea arabica* L. Coffee (Turrialba) 3:97–115

Muller CJ, Jennings WG (1967) Constituents of black pepper. Some sesquiterpene hydro- carbons. J Agric Food Chem 15:762–766

Muller CJ, Creveling RK, Jennings WG (1968) Some minor sesquiterpene hydrocarbons of black pepper. J Agric Food Chem 16:113–117

Murashige T, Skoog F (1962) A revised medium for rapid growth and bioassays with tobacco tissue culture. Physiol Plant 15:473–497

Nazeem PA, Joseph L, Thampi MS, Sujatha R, Nair GS (1993). *In vitro* culture system for indirect organogenesis for black pepper (*Piper nigrum* L.) In: Golden Jubilee symposium on horticultural research-changing scenario, 24–28 May, Bangalore, India, p 250

Nigam MD, Handa KL (1964) Constituents of essential oils. Indian Perfumer 8:15–17

Nirmal Babu K, Naik VG, Ravidran PN (1993a) Two new taxa of Piper (*Piperaceae*) from Kerala, India with a note on their origin and inter-relationships. J Spices Aromatic Crops 2:26–33

Nirmal Babu K, Rema J, Lukose R, Ravindran PN, Johnson George K, Sasikumar B, Peter KV (1993b) Spices biotechnology at National Research Centre for Spices, Technical Report II, Calicut, Kerala State, India

Nirmal Babu K, Geetha SP, Manjula C, Sajina A, Minoo D, Samsudeen K, Ravindran PN, Peter KV (1996) Biotechnology – its role in conservation of genetic resources of spices. In: Das MR, Mundayoor S (eds) Biotechnology for development. State Committee on Science, Technology and Environment, Trivandrum, pp 198–212

Pangburn RM, Jennings WG, Noelting CF (1970) Preliminary examination of odour quality of black pepper oil. Flavour Ind 1:763
Parmar VS, Jain SC, Bisht KS, Jain R, Taneja P, Jha A, Tyagi OD, Prasad AK, Wengel J, Olsen CE, Boll PM (1977) Phytochemistry of the genus *Piper*. Phytochemistry 46:597–673
Philip VJ, Joseph D, Trigs GS, Dickinson NM (1992) Micropropagation of black pepper (*Piper nigrum* L.) through shoot tip cultures. Plant Cell Rep 12:41–44
Pruthi JS, Bhat AV, Kutty SK, Varkey AG, Gopalakrishnan M (1976) Preservation of fresh green pepper by canning, bottling and other methods. International seminar on pepper. Spices Export Promotion Council of India, Cochin, pp 64–72
Purseglove JW, Brown EG, Green CL, Robbins SRJ (1981) Spices, vol 1. Longman, London, pp 1–99
Rahiman BA (1981). Biosystematic studies in varieties and spices of *Piper* occurring in Karnataka region. Ph.D. thesis, University of Mysore, Karnataka State, India
Raj HG (1978) A comparison of the systems of cultivation of Black pepper *Piper nigrum* L. in Malaysia and Indonesia. The Directorate of Extension Education, Kerala Agricultural University, Trichur, Kerala State, India, pp 65–75
Raj Mohan K (1985) Standardization of tissue culture techniques in important horticultural crops. Ph.D. thesis, Kerala Agricultural University, Thrissur, Kerala State, India
Ramadasan A (1987) Canopy development and yield of adult pepper vines in relation to light interception. Indian Cocoa Arecanut Spices J 11:43–44
Ravindran PN (1991) Studies on black pepper and some of its wild relatives. Ph.D. thesis, University of Calicut, Calicut, Kerala State, India
Ravindran PN, Nirmal Babu K (1994) Genetic resources of black pepper. In: Chadha KL, Rethinam P (eds) Advances in Horticulture. Malhotra Publications Co., New Delhi, pp 99–120
Ravindran PN, Peter KV (1995) Biodiversity of major spices and their conservation in India. In: Arora RK, Rao VR (eds) Proceedings of the South Asia Coordinators meeting on plant genetic resources. IPGRI, New Delhi, pp 123–134
Ravindran PN, Nair MK, Nair RA (1987) Two taxa of *Piper* (Piperaceae) from Silent Valley Forests, Kerala. J Econ Tax Bot 10:167–169
Ravindran PN, Nair MK, Nair RA, Nirmal Babu K, Chandran K (1990) Ecological and taxonomical notes on Piper spp. from Silent Valley Forests, Kerala. J Bombay Nat His Soc 87:421–426
Richard HM, Jennings WG (1971) Volatile composition of black pepper. J Food Sci 36:584–589
Richard HM, Russel GF, Jennings WG (1971) The volatile components of black pepper varieties. J Chromatogr Sci 9:560–566
Rosengarten F Jr (1973) The book of spices. Pyramid Books, New York
Russel GF, Else J (1973) Volatile compositional differences between cultivars of black pepper (*Piper nigrum*). J Assoc off Anal Chem 56:344–351
Sajina A, Minoo D, Geetha SP, Samsudeen K, Rema J, Nirmal Babu K, Ravindran PN, Peter KV (1996) Production of synthetic seeds in spices. In: Edison S, Ramana KV, Sasikumar B, Nirmal Babu K, Eapen SJ (eds) Biotechnology of spices, medicinal and aromatic plants. Indian Society for Spices, Calicut, pp 65–69
Salzer UJ (1975) Uber die Fettsaurezussamensetzung de lipoide einiger gewurze. Fette Stefen Anstrichm. 77:446–450
Sasikumar B, Veluthambi K (1994) Kanamycin sensitivity of cultured tissues of *Piper nigrum* L. J Spices Aromatic Crops 3:158–160
Schenk RU, Hildebrandt AC (1972) Medium and techniques for induction and growth of monocotyledonous and dicotyledonous plant cell cultures. Can J Bot 50:199–204
Sharma NL, Nigam NC, Handa KL (1962) Pepper husk in oil. Perfuem Kosmet 43:505–506
Sim SL, Jafar R, Grierson D, Power JB, Davery MR (1995) Applications of molecular biology and genetic manipulation to pepper (*Piper nigrum* L.) breeding: present state and prospects. In: Abstracts FAO/IAEA international symposium on the use of induced mutations and molecular techniques for crop improvement, Vienna, Austria

Trelease W, Yuncker TG (1950) The *Piperaceae* of Northern South America. University of Illinois, Urbana

Velayudhan KC, Amalraj VA (1992) *Piper pseudonigrum*—a new species from Western Ghats. J Econ Tax Bot 16:247–250

Vijayakumar KR, Unni PN, Vamadevan VK (1984) Prevention of photo-induced chloro- phyll loss by the use of lime reflectant on the leaves of black pepper (*Piper nigrum* L.). Agric For Meteorol 34:17–20

Wealth of India (1969) *Piper nigrum* Linn (Piperaceae) Wealth of India – raw materials, vol 8. Publications and Information Directorate, Council of Scientific and Industrial Research, New Delhi, pp 99–115

Wrolstad RE, Jennings WG (1965) Volatile constituents of black pepper III. The monoter- pene hydrocarbon fraction. J Food Sci. 30:274–279

Yuncker TG (1958) The Piperaceae: a family profile. Brittonia 10(1):17

Chapter 3
Pepper Agronomy

Abstract The chapter describes suitable pepper soils, climatic conditions, and agronomic practices. There is an extensive discussion in the chapter on the nutrition of black pepper, the evolution of pepper manuring, and the response of pepper to mineral nutrition, as well as discussions on specific nutrient deficiency symptoms, with pictorial representations.

Keywords Agronomy · Soils · Pepper manuring

Agriculture systems differ from natural systems in one fundamental aspect: while there is a net outflow of nutrients by crop harvests from soils in the first, there is no such thing in the second (Sanchez 1994). This is because nutrient losses due to physical effects of soil and water erosion are continually replenished by weathering of primary minerals or atmospheric deposition. Hence, the crucial element of sustainability is the nutrient factor. Of all the factors, however, the nutrient factor is the least resilient (Fresco and Kroonenberg 1992). The thrust of high-input technology—the hallmark of the green revolution, in retrospect—or the moderation by low-input technology, the foundation stone of sustainable agriculture, in prospect, both dwell on this least resilient nutrient factor. If the pool of nutrients in the soil, both native and added, could be considered as the "capital," efficient nutrient management might be analogous to raising the "interest" accrued from this capital in such a way that there is no great danger of erosion of this capital. Hence, sound prescriptive soil management should aim at understanding the actual link between the capital and the interest so that meaningful practices can be prescribed (Nair 1996). This is all the more important for a perennial crop like pepper whose life span can be upward of 25 years. An effort in this direction was initiated recently by Nair (2002). But, before we dwell on it, it is only fair that an objective review is made of what has been done so far on the agronomy of black pepper.

K. P. Nair, *The Geography of Black Pepper (*Piper nigrum*)*,
https://doi.org/10.1007/978-3-030-52865-2_3

3.1 The Pepper Soils

Pepper easily grows in a variety of soils. De Waard (1969) reported that in Malaysia, pepper mostly grows on soils that are developed from slate or sandstone and even in soils of alluvial origin low in fertility. Pepper is seen to grow in Indonesia in all types of soils—ranging from fertile and friable volcanic soils to sticky clayey soils. In Sri Lanka, pepper growers prefer red clay loam or sandy loam. In India, pepper is seen to grow in a variety of soils, but, in its native Kerala State, it is grown mostly in red laterites. Pepper prefers a well-drained soil having adequate water-holding capacity and that is rich in humus and essential plant nutrients. In India, the major pepper-growing soils are oxisols (6%), alfisols (70%), mollisols (10%), and entisols (4%). The statewise distribution is outlined in (Table 3.1) (Sadanandan 1994).

Forest loam is acidic in reaction, with a pH range of 5.0–5.5, and is confined mainly to the Western Ghats Highlands. The soils are rich in organic carbon, well-drained, brown to black in color, and very well suited for pepper. Soils, in general, are quite fertile, rich in nitrogen and potassium, and medium in phosphorus. Laterites are, again, acidic, with a pH range of 5.0–6.2, low in fertility, and invariably run into problems with phosphorus fertilization because of high amounts of soluble iron and aluminum, which render applied phosphorus immobile. Excessive soluble aluminum leads to aluminum toxicity. Of late, micronutrient deficiencies—in particular, zinc—have been reported, and recently Nair (2002) demonstrated the lacunae in zinc nutrition of pepper based on classical textbook knowledge of zinc fertilization in these problematic soils. Alluvium is confined to the riverbanks and their tributaries. The soils are only moderately fertile and respond well to management. The soils are acidic in nature, with pH ranging from 5.0 to 6.5. Red loams are highly porous and friable, low in fertility, and, like laterites, contain high amounts of soluble aluminum and iron, which lead to problems with phosphorus fertilization. Except in forest loam, where pepper is grown as a monocrop because the soil is highly suitable for the crop, in all the others it is grown both singly or in association with other crops, such as coconut (*Cocos nucifera*), areca nut (*Areca catechu*), and the like. Unlike the large exclusive pepper plantations, these soils support what is generally known as "homestead farming" in Kerala—small, less than 5 ha, self-supporting family farming—where even milch animals are reared in association.

Table 3.1 Type, order, and suborder of major pepper-growing soils and their statewise distribution in India

Soil type	Order	Suborder	Indian state
Forest loam	Mollisols	Udolls	Kerala, Karnataka, and Tamil Nadu
		Ustolls	
Laterite	Alfisols	Ustalfs	Kerala, Karnataka, and Tamil Nadu
	Oxisols		
Alluvium	Entisols	Ustorthcut	Kerala and Karnataka
Red loam	Alfisols	Ustalfs	Kerala and Karnataka
		Ustults	

3.2 Nutrition of Black Pepper

The fundamental difference between agricultural systems and natural systems is that while there is a net outflow of plant nutrients by crop harvests from soils in the former, there is no such thing in the latter. This is because nutrient losses due to physical effects of soil and water erosion are continually replenished by weathering of primary minerals or atmospheric deposition in natural systems. Hence, the crucial element of sustainability of crop production is the nutrient factor. As discussed earlier, the nutrient factor is the least resilient to management. For production sustainability, the nutrient factor is the most crucial. It is becoming increasingly clear that it is not the quantum per se of a specific nutrient in the soil that is crucial, as far as plant requirements are concerned, but its bioavailability. In the final analysis, it is the plant and plant alone that will decide whether a nutrient inherently present in the soil or externally added to it will ultimately be "available" to the plant or not (Nair 2002). Formidable effort has gone into the task of defining this availability. It is the conviction of this author that if sound and accountable prescriptive fertilizer recommendations have to be made to sustain crop production in the decades to come, one must have a clear understanding of the dynamics of nutrient availability from which accurate predictions can be made and sustainable and accountable field practices devised. These considerations are particularly important to devise accurate fertilizer recommendations for pepper because, unlike most other annual or biennial crops, pepper is perennial, and the utilization pattern of applied or native nutrients over several years, often running into decades, could be uniquely different. This part of the review will, at first, dwell on the available pool of information relating to the general and prevalent mode of pepper nutrition and then will go on to the next stage, where it will encompass the newer vision, which is now known, the world over, as "the nutrient buffer power concept."

3.3 Evolution of Pepper Manuring

Pepper has been in cultivation for many decades in India and elsewhere. In earlier days, farmers used only meager quantities of organic manures, such as leaf litter, animal manure (principally cow dung, which is a widespread practice of homestead farming in Kerala even now), or slashed stems and leaves of live support, such as *Erythrina indica*. Chemical manuring, through factory-manufactured fertilizers, is a relatively recent phenomenon. It was the pepper boom—and the escalating prices of pepper in the world market, especially in the post-World War II period—that encouraged pepper farmers to use artificial manure. In addition, the spectacular initial yield increases in wheat and rice in the North Indian regions—the green revolution effect owing to liberal doses of chemical fertilizers—prompted affluent pepper farmers to use factory-produced fertilizers. Though there is now a backlash against the use for chemical fertilizers—in terms of degraded soils, drying aquifers, and vanishing biodiversity, all of which are attributed to the indiscriminate use of chemical fertilizers—chemical fertilizer use in a systematic manner is still confined only to affluent owners of pepper plantations. Organic manuring is still the mode of pepper nutrition in homestead farming. Significantly, even pepper planters of fairly large farms are gradually switching to organic farming, because "organic pepper" raised by the use of only organic manures fetches a much better price, and a steady and premium market is developing, especially in European countries such as Germany, where the spices are much fancied lately.

3.4 Response of Pepper to Mineral Nutrients

The major emphasis on pepper fertilization is still confined to nitrogen, phosphorus, and potassium. There is an extensive body of literature on nitrogen fertilization in pepper globally (De Waard 1964; Mohanakumaran and Cheeran 1981; Pillai and Sasikumaran 1976; Pillai et al. 1979, 1987; Sadanandan 1994; Sim 1971). These routine studies have dealt with partitioning of nitrogen in various plant parts (Adzemi et al. 1993), response functions (Pillai et al. 1987; Sadanandan 1990), greenhouse cultivation (Murni and Faodji 1990), and foliar nitrogen application (Anon 1995). Adzemi et al. (1993) found the maximum nitrogen concentration in leaves (2.30%), while the branches transported the most (47.6 mg plant^{-1} year^{-1}). In Kerala State, for Panniyur-1, the first hybrid variety released in the world from the Pepper Research Station, Panniyur, 50 kg nitrogen with 100 kg P_2O_5 and 150 kg ha^{-1} K_2O were found to be the optimum rates for laterite soil (Pillai et al. 1987). In greenhouse conditions, 1.1 g plant^{-1} nitrogen as urea and 2.0 g plant^{-1} K_2O produced the maximum of dry matter (Murni and Faodji 1990). Between Panniyur-1 and Karimunda, another improved variety from the Panniyur Research Station, 292 kg ha^{-1} year^{-1} and 183 kg ha^{-1} year^{-1} of nitrogen, respectively, was found to be the optimum rates (Sadanandan 1990). In Sarawak (Malaysia), foliar application of 0.7% urea at weekly intervals, totaling nine sprays, was found to increase pepper yield by 22% (Anon 1995) (Table 3.2).

Comparatively, research on phosphorus in pepper is less, viewed against that of nitrogen. Unlike nitrogen, pepper stems contain the most phosphorus (3.96 g vine^{-1}) because of large transport, and a total of 22.8 kg ha^{-1} of phosphorus is removed from the soil (Adzemi et al. 1993). Availability of phosphorus in laterite soil is a universal problem, because of the presence of excessive soluble aluminum and iron, which renders the applied phosphorus fertilizer immobile, and some researchers have focused on alternate sources of phosphorus as a possible remedy. Phosphate rocks (PRs) are ideally suited for plantation crops and where indigenously available could be profitably utilized. In northern India, a popular PR is Mussoorie rock phosphate (MRP). Sadanandan (1994) reported that MRP is superior to ordinary superphosphate. Both cumulative yield increase and relative agronomic effectiveness were superior in MRP as compared to ordinary superphosphate.

Table 3.2 Nutrient deficiency symptoms in pepper

Nutrient	Deficiency symptoms	Cited by
Nitrogen	General chlorosis and stunted growth. Yellowing of older leaves followed by younger ones. Bottom leaf tips and margins become brown and necrotic	De Waard (1969) and Nybe and Nair (1986)
Phosphorus	Bronzing of older leaves accompanied by necrosis of leaf tips and margins. Stunted growth	De Waard (1969) and Nybe and Nair (1986)
Potassium	Browning and necrosis of older leaf tips and margins. Symptoms later spread to younger leaves	De Waard (1969) and Nybe and Nair (1986)
Calcium	Young leaves develop tiny, brown necrotic pinhead spots that later spread to older leaves. Interveinal chlorosis can also be seen	De Waard (1969) and Nybe and Nair (1986)
Magnesium	Pale yellow discoloration of leaf margins and tips, followed by necrosis and defoliation. Major veins remain green, and laterals turn yellow	De Waard (1969) and Nybe and Nair (1986)
Sulfur	Late stage chlorosis in younger leaves, turning to bright yellow color in interveinal areas. Premature leaf fall and die-back of growing tip	Nybe and Nair (1986)
Zinc	Stunted growth, small leaves, interveinal chlorosis. Leaf margins pucker	Nybe and Nair (1986)
Manganese	Interveinal chlorosis, with major veins remaining green. Chlorotic leaves turn yellow or white later and necrotic mature leaves. Symptoms can often be confused with that of magnesium deficiency	Nybe and Nair (1986)
Iron	Interveinal chlorosis in younger leaves, youngest leaves becoming totally chlorotic	Nybe and Nair (1986)
Copper	Interveinal chlorosis in younger leaves; necrosis on leaf tips and margins	Nybe and Nair (1986)
Boron	Stunted growth, necrosis, and interveinal chlorosis. Necrotic lesions are seen on main vein	Nybe and Nair (1986)

Pepper is a prolific user of potassium, and even with 2% content of potassium in the pepper plant, potassium deficiency will manifest (De Waard 1969). Field trials conducted using Panniyur-1 as the test crop showed that as much as 200 g of K_2O vine^{-1} is required (Pillai et al. 1987). Response functions to potassium fertilizer application indicated that pepper needs a very high dose of potassium (270 kg K_2O ha^{1}) for high yield (Sadanandan 1990). Pepper plants remove as much as 203.2 kg ha^{-1} of potassium, and white pepper stores most of it (42.0 g vine^{-1}) as reported by Adzemi et al. (1993). Most pepper growers adopt varying nitrogen-to- phosphorus-to-potassium (N:P:K) ratios, while Pillai et al. (1979) suggest an optimum ratio of 5:5:10 for nitrogen, phosphorus, and potassium to obtain the maximum yield.

As far as the secondary nutrients (calcium, magnesium, and sulfur) are concerned, the only published work refers to calcium indirectly, through the effect of liming. Pillai et al. (1979), however, did not observe any positive response to lime application in a laterite soil growing Panniyur-1. Adzemi et al. (1993) report the highest accumulation of both calcium and magnesium in stems (11.5 and 5.98 g vine^{-1}, respectively), with a total accumulation of 54.5 and 36.4 kg ha^{-1}, respectively.

Among the micronutrients, zinc is the most important followed by molybdenum and boron (De Waard 1969). Zinc deficiency is beginning to be seen in many tropical countries in a very significant manner and is adversely affecting crop yields. However, to date, corrective measures still bank on routine soil testing with DTPA as the most commonly used extractant. That such an approach does not lead to satisfactory results uniformly is shown through the work of this author (Nair 2002), which will be discussed later in this section. A combination of zinc, molybdenum, and boron, along with nitrogen, phosphorus, and potassium, was shown to result in the highest yield of two locally developed varieties—namely, Subhakara and Sreekara at the Indian Institute of Spices Research experimental farm (IISR 1997), where zinc, boron, and molybdenum were used at a ratio of 5:2:1. Spraying a solution of $ZnSO_4$ (0.5%) reduced spike (inflorescence which eventually develops into pepper berries) shedding by 48.4% (Geetha and Nair 1990).

Organic manuring is a very commonly adopted practice in pepper production in India and parts of Asia. This can either be through the use of fresh vegetative matter or through the use of "burnt earth" (Bergman 1940; Harden and White 1934). In the former category, freshly chopped materials (such as leaves, stems, and the like) from a number of trees are used. The trees generally used are *E. indica*, *Garuga pinnata*, and *Grevillea robusta* (Sivakumar and Wahid 1994). Organic manures are widely used in pepper production in Sarawak (Malaysia), which include soybean cake, guano, prawn, and fish refuse (Purseglove et al. 1981). Holes are dug, and about 85 g of manure are placed about 20 cm away from the main stem of the vine. This is done once in a 2-month period. Some farmers dig trenches all around the vines and put large quantities of manure once in 2 months. Of late, sterilized animal meat and bone meal admixture, fortified with potassium, are gaining popularity as organic manure.

Another method of organic manuring is through the use of "burnt earth." For this, forests are first cleared; then, soil from vacant patches is spread on top of the heaped vegetation, the heap is set on fire, and the fire is allowed to slowly burn for 2–3 weeks. At the end of this period, both the wood ash and burnt soil are thoroughly mixed and applied at a rate of 18 kg vine^{-1} year^{-1}. The application of burnt ash has many beneficial effects in improving the physical, chemical, and biological properties of soil. Compared to the acidic soil, having a pH of 4–5, the burnt earth has a pH of 7–8 and acts as a good soil ameliorant in correcting pH. There is extensive literature on this method (Bergman 1940; De Waard 1969; Harden and White 1934; Huitema 1941). Owing to environmental hazards, especially because of large-scale burning and the emission of smoke, the Government of Sarawak banned the practice in 1940. Figures 3.1, 3.2, and 3.3 represent potassium, magnesium, and zinc deficiency, respectively.

Fig. 3.1 Potassium deficiency symptom

Fig. 3.2 Magnesium deficiency symptom

Fig. 3.3 Zinc deficiency symptom

References

Adzemi MA, Hamdan J, Sjahril JS (1993) Micronutrient removal studies on black pepper (*Piper nigrum* L.) in Sarawak. In: Ibrahim MY, Bong CFJ, Ipor IP (eds) The pepper industry: problems and prospects. Universiti Pertanian Malaysia, Sarawak, pp 104–109

Anon (1995) Annual report of the Research Branch, Department of Agriculture, Sarawak, Malaysia

Bergman A (1940) De pepper culture en-handel op Bangka. Landbouw 16:139–256

De Waard PWF (1964) Pepper cultivation in Sarawak. World Crops 16:24–30

De Waard PWF (1969) Foliar diagnosis, nutrition and yield stability of black pepper (*Piper nigrum* L) in Sarawak. Royal Tropical Institute, Amsterdam, The Netherlands, p 150. Comm. No. 58

Fresco LO, Kroonenberg SB (1992) Time and spatial scales in ecological sustainability. Land Use Policy:155–167

Geetha CK, Nair PCS (1990) Effect of plant growth regulators and zinc on spike shedding and quality of pepper. Agric Res J Kerala 14:10–12

Harden HJ, White JTH (1934). Quoted by De Waard (1969)

Huitema WK (1941) Quoted from De Waard (1969)

IISR (1997) Annual Report of Indian Institute of Spices Research, 1996–97, Calicut, Kerala State, India, pp 50–51

Mohanakumaran B, Cheeran A (1981) Nutritional requirement of pepper vine trailed on living and dead standards. Agric Res J Kerala 19:3–5

Murni AM, Faodji R (1990) The combination effect of KCl with two nitrogen sources on the growth of black pepper. Bull Penelitian Tanaman Rempah dan Obat 2(2):79–84

Nair KPP (1996) The buffering power of plant nutrients and effects on availability. Adv Agron 57:237–287

Nair KPP (2002) Sustaining crop production in the developing world through the Nutrient Buffer Power Concept. In: Proceedings of the 17th World Soil Science Congress, Bangkok, Thailand, August 14–21, vol. 2, p 652

Nybe EV, Nair PCS (1986) Nutrient deficiency in black pepper (*Piper nigrum* L.) I. Nitrogen, Phosphorus and Potassium. Agric Res J Kerala 24:132–150

Pillai VS, Sasikumaran S (1976) A note on the chemical composition of pepper plant (*Piper nigrum* L.). Arecanut Spices Bull 8:13–17

Pillai VS, Chandy KC, Sasikumaran S, Nambiar PKV (1979) Response of Panniyur-1 variety to nitrogen and lime application. Indian Cocoa Arecanut Spices J 3(2):35–38

Pillai VS, Sasikumaran S, Nambiar PKV (1987) N, P and K requirement of black pepper. Agric Res J Kerala 25(1):74–80

Purseglove JW, Brown EG, Green CL, Robbins SRJ (1981) Spices, vol 1. Longman, London, pp 1–99

Sadanandan AK (1990) Response of pepper to applied potassium (Abs.). National Seminar on Potassium on Plantation Crops, Bangalore, India, p 149

Sadanandan AK (1994) Nutrition of black pepper. In: Chadha KL, Rethinam P (eds) Advances in horticulture. Vol 9: plantation and spice crops (Part I), New Delhi, pp 423–456

Sanchez PA (1994) Tropical soil fertility research: towards the second paradigm. In: Proceedings of the XV ISSUS Symposium, Acapulco, Mexico, July 10–16, vol 1, pp 89–104

Sim ES (1971) Dry matter production and major nutrient contents of black pepper (*Piper nigrum* L.). Malay Agric J 48:73

Sivakumar C, Wahid PA (1994) Effect of application of organic materials on growth and foliar nutrient contents of black pepper (*Piper nigrum* L.). J Spices Aromatic Crops 3:135–141

Chapter 4
The Role of the Nutrient Buffer Power Concept in Pepper Nutrition

Abstract The chapter describes the role of a revolutionary soil management technique, globally known as "The Nutrient Buffer Power Concept" in pepper nutrition. It develops the basic thermodynamics of plant nutrient bioavailability using mathematical models, based on the nutrient diffusive process in the soil, and goes on to precisely quantify the most important parameter which will define the bioavailability of crucial nutrients such as zinc. The author of this concept mathematically defines the "buffer power" of the crucial nutrient in black pepper, such as zinc.

Keywords Buffer power · Diffusion · Plant nutrition · Zn Buffer power

Historically, soil testing has been used to quantify availability of essential plant nutrients to field-grown crops. However, contemporary soil tests are based on philosophies and procedures developed several decades ago without significant changes in their general approach. For a soil test to be accurate, one needs to clearly understand the physico-chemico-physiologic processes at the soil-root interface, and an understanding of soils and plant root systems as polycationic systems is as essential. It is this knowledge that leads to sound prescriptive soil management practices in nutrient availability vis-à-vis fertilizer application, because, of all the factors that govern sustainability of crop production, the nutrient factor is the most important, and it is also the least resilient to management. This section of the review will focus on the buffering of plant nutrients, with specific reference to zinc, and discuss experimental results that relate to pepper nutrition.

K. P. Nair, *The Geography of Black Pepper (*Piper nigrum*)*,
https://doi.org/10.1007/978-3-030-52865-2_4

4.1 The Buffer Power and Its Effects on Nutrient Availability

Before being able to understand the significance of the *nutrient buffer power concept* on plant nutrient availability, certain basic concepts must be addressed, and the following review starts with this rationale.

4.2 Basic Concepts

In any nutrient management approach that is sound and reproducible, one must start with a basic understanding of the chemical environment of plant roots. When we consider this, the first term that we come across is the *soil solution* because the plant root is bathed in it, and it is most affected by its chemical properties. The Soil Science Society of America (1965) defines soil solution as "the aqueous liquid phase of the soil and its solutes consisting of ions dissociated from the surfaces of the soil particles and of other soluble materials." Adams (1971) has given a simple definition: "The soil solution is the aqueous component of a soil at field moisture content." Perhaps it is important to emphasize here that much contemporary soil testing has considered a soil extract as synonymous with the soil solution. Because soil extraction is supposed to simulate plant root extraction, it is pertinent to consider the chemical environment of the root, though briefly, from this angle. It is worth noting that the chemical environment of roots in natural soil systems is so obviously complex that both soil scientists and plant physiologists have been unable to provide a precise definition. If this complex chemical system is to be accurately quantified, thermodynamic principles will need to be used to evaluate experimental data. Even then, the limitations are obvious, as in the case of potassium, where the thermodynamic investigations are quite often inapplicable under field conditions. Thus, although a quasi-equilibrium in potassium exchange can be achieved in the laboratory, these conditions are seldom, if ever, attained under field conditions (Sparks 1987). Agricultural soils are, for the most part, in a state of disequilibrium owing to both fertilizer input and nutrient uptake by plant root. It appears that a universal and accurate definition of a root's chemical environment awaits the proper application of thermodynamics for the root's ambient solution (Adams 1971) or even kinetics, as in the case of potassium (Sparks 1987), where thermodynamics have been found to be inadequate.

Soil extractions with different extractants provide a second approach in defining the root's chemical environment. This approach has been particularly successful in understanding cases like phosphorus insolubility, soil acidity, and potassium fixation. However, this approach also fails to precisely define the root's chemical environment. Though this approach also suffers deficiencies—such as the extractants removing arbitrary and undetermined amounts of solid-phase electrolytes and iron (or the extractants causing precipitation of salts or ions from the soil solution) and the soil-plant interrelationship defined in terms of solid-phase component of the soil, even though the solid phase is essentially inert except as it maintains thermodynamic equilibria with the solution phase (Adams 1971)—the latter part could be researched more to understand how the solid phase-solution phase equilibria can be interpreted to give a newer meaning to quantifying nutrient availability. It is in this context that the role of the plant nutrient *buffer power* assumes crucial importance.

The close, almost linear relationship in a low concentration range of <0.5 mM for NO_3^-N, NH_4^+–N, K^+, $H_2PO_4^-$, HPO_4^{2-} which has been established by numerous solution culture experiments, can be quantitatively described by the equation:

$$U = 2\pi\, r\alpha C_r \tag{4.1}$$

where U is the uptake of a 1 m root segment, r is the root radius, C_r is the concentration of the ion at the root surface, and α is the root-absorbing power (Mengel 1985). The metabolic rate of the root determines its absorbing power. A high root-absorbing power would imply that a relatively high proportion of nutrient ions coming in contact with the root surface is absorbed and vice versa. The nutrient ion concentration at the root surface (C_r) depends on α because a high root-absorbing power tends to decrease C_r; it also depends on the rate of movement from bulk soil toward the root (Mengel 1985). Diffusion and/or mass flow control this movement. But it is now established that nearly 95% of this movement for nutrients such as phosphorus, potassium, and zinc (among heavy metals) and, possibly, NH_4 is by way of diffusion. When root uptake of an ion species is less than its movement toward it, accumulation of the ion species on the root surface is bound to occur, as has been shown to be the case with Ca^{2+}, where mass flow contributes to this accumulation (Barber 1995). The diffusive path for ions such as phosphorus and potassium, which plant roots take up at high rates but which are in low concentration in the soil solution near the root, is the concentration gradient. In a sense, the effective diffusion coefficient that quantifies the diffusive path and the buffer power are analogous because the diffusive flux across the root surface is integrally related to the nutrient buffer power. This has been shown to be true in the case of phosphorus, where a highly significant positive correlation between the two was found to exist in 33 soil samples obtained from experimental sites in the USA and Canada (Kovar and Barber 1988). However, in a routine laboratory setup, it is far easier to measure the buffer power than the effective diffusion coefficient, and this review will further focus on the question of how buffer power can be quantified without recourse to cumbersome analytical techniques and how its integration into routine soil test data will considerably improve predictability of nutrient uptake.

4.3 Measuring the Nutrient Buffer Power and Its Importance in Affecting Nutrient Concentration on Root Surfaces

The ability to predict the mobility of dissolved chemicals, such as fertilizers, in the soil is of considerable value in managing fertilizer applications. Soil testing in its essence aims to achieve this. While modeling transport and retention of ions from thermodynamic (Selim 1992), kinetic (Sparks 1989), and mechanistic (Barber 1984) angles could be very informative, the importance of translating this information into practically feasible procedures in crop production calls for an understanding not only of the basic concepts but of their intelligent application as well. In a dynamic state of plant growth, the concentration of any nutrient on the root surface is nearly impossible to measure because both the nutrient in the plant tissue and the root-absorbing power, which directly affects it, change quickly due to root metabolic processes. The inability of even mechanical mathematical models to accurately predict the nutrient influx rate has recently been highlighted (Lu and Miller 1994). Hence, if an effective soil testing procedure is to be devised for a nutrient, which is an alternative to defining the plant root's chemical environment, one must resolve the problems of quantifying the nutrient concentration on the root surface indirectly, even if it is impossible to resolve it directly, for the reasons mentioned earlier. Using Fick's first law:

$$F = -D\left(\frac{dC}{dx}\right) \tag{4.2}$$

where F is the flux, dC/dx is the concentration gradient across a particular section and D is the diffusion coefficient. Nye (1979) has suggested that the formula can be applied to both ions and molecules. The negative sign for D implies net movement from a high to a low concentration. Although, for molecules in simple systems, like dilute solutions D may be nearly constant over a range of concentrations, for ions in complex systems like soils and clays, D will usually depend on the concentration of the ion and on that of other ions as well (Nye 1979). Nye has further suggested that though Fick's first law may be derived from thermodynamic principles in ideal systems, in a complex medium such as the soil, Equation (4.2) may be regarded as giving an operational definition of the diffusion coefficient. Thus, Nye (1979) defines the diffusion coefficient as

$$D = D_1 \theta f_1 \frac{(dC_1)}{dC} + DE \tag{4.3}$$

where D_1 is the diffusion coefficient of the solute in free solution, θ is the fraction of the soil volume occupied by solution and gives the cross-section for diffusion, f_1 is an impedance factor, C_1 is the concentration of solute in the soil solution, and D_E is an excess term which is zero when the ions or molecules on the solid have no surface

Table 4.1 Comparison of phosphorus buffer power of eight widely differing central European soils (determined by two different techniques)

	Regression		r	
Soil	(1)	(2)	(1)	(2)
Benzheimer Hof	$Y = 18.8x + 7.94$	$Y = 0.23x + 8.98$	0.912	0.995
Hungen	$Y = 38.2x - 1.03$	$Y = 0.25x$ 4.32	0.967	0.997
Oldenburg B6	$Y = 49.8x + 0.52$	$Y = 0.26x+ 0.72$	0.994	0.999
Wolfersheim	$Y = 70.3x + 0.03$	$Y = 0.27x + 0.11$	0.998	0.983
Obertshausen	$Y = 70.5x + 2.66$	$Y = 0.30x + 2.89$	0.966	0.998
Oldenburg B3	$Y = 73.6x + 2.07$	$Y = 0.31x + 0.61$	0.994	0.997
Klein-Linden	$Y = 75.0x + 0.38$	$Y= 0.32x + 1.81$	0.999	0.991
Gruningen	$Y = 75.4x + 0.89$	$Y = 0.36x + 3.62$	1.000	0.996

Note: The b values in the regression functions represent the phosphorus buffer power of each soil. In regression function
(1) After Nair and Mengel (1984), $Y =$ CAL-P (Schüller's method), and in regression function (2) After Nair (1992), $Y =$ the author's method, x in both refers to electroultrafiltrable phosphorus. Note the very high r values in all the cases. The soils are arranged in their sequential increase in phosphorus buffer power

mobility but represents their extra contribution to the diffusion coefficient when they are mobile. D_E can generally be neglected because only in rare instances will it play any role in diffusion of plant nutrient ions in soil (Mengel 1985). From the point of view of nutrient availability, dC_1/dC, which represents the concentration gradient, assumes crucial importance as we shall see next.

The term dC_1/dC, where C_1 is the concentration of the nutrient ion in the soil solution and C is the concentration of the same ion species in the entire soil mass, assumes considerable significance in lending a practical meaning to nutrient availability. If we ascribe the term "capacity" or "quantity" to C and "intensity" to C_1, we have in this term an integral relationship between two parameters that may crucially affect nutrient availability. Because the concentration gradient of the depletion profile of the nutrient in the zone of nutrient uptake depends on the concentration of the ion species in the entire soil mass (represented by "capacity" or "quantity") in relation to the rate at which this is lowered on the plant root surface by uptake (represented by "intensity"), it could be argued that a quantitative relationship between the two should represent the rate at which nutrient depletion and/or replenishment in the rooting zone should occur (Nair 1984a). This relationship has been functionally quantified by Nair and Mengel (1984) for phosphorus in eight widely differing Central European soils (Table 4.1), and the term dC_1/dC has been referred to as the nutrient buffer power. Nair and Mengel (1984) used electroultrafiltration to quantify C_1, while using an incubation and extraction technique to quantify C. For phosphorus, C was found to closely approximate isotopically exchangeable phosphorus (Keerthisinghe and Mengel 1979), but in the experiments conducted by Nair and Mengel (1984), it was estimated by the extraction of incubated soil with an extractant that was a mixture consisting of 0.1 M calcium lactate + 0.1 M calcium acetate + 0.3 M acetic acid at pH 4.1. The extractant exchanges adsorbed phosphate and dissolves calcium phosphates except apatites; the method known as the "CAL method," developed by Schüller (1969), is now widely used in Central Europe. In the case of K^+ and NH_4-N, C denotes the concentration of exchangeable, and to some extent nonexchangeable, fractions (Mengel 1985). Because very low concentrations in the range of 2.0 μM may be attained on the root surface for both phosphorus and potassium (Claassen and Barber 1976; Claassen et al. 1981; Hendriks et al. 1981), Nair and Mengel (1984) had to use electroultrafiltration to quantify C_1.

Thus, the nutrient depletion around the roots that is caused by the diffusive flux of the nutrient toward the root surface is related to both the quantity and the intensity parameters, and a quantifiable relationship between both represents the buffer power specific to the nutrient and the soil. A growing root will at first encounter a relatively high concentration of phosphorus, which is in the range of the concentration of the bulk soil solution (Nair and Mengel 1984). As uptake continues, depletion will occur at the root surface. This depletion profile gets flatter with enhanced nutrient uptake (Claassen et al. 1981; Hendriks et al. 1981; Lewis and Quirk 1967). But it is the capacity of the soil to replenish this depletion which ensures a supply of nutrient ions to the plant root without greatly depressing its average concentration on the root surface. It is the nutrient's buffer power that decides these depletion and/or replenishment rates. A soil with a high phosphorus buffer power implies that the phosphorus absorbed from the soil solution is rapidly replenished. In such a case, phosphorus concentration at the root surface decreases only slowly, and mean phosphorus concentration at the root surface remains relatively high. In soils with a low phosphorus buffer power, the reverse is true, and mean phosphorus concentration at the root surface is rapidly diminished and remains relatively low. This has been proved experimentally for phosphorus (Nair 1992; Nair and Mengel 1984). This phenomenon holds true for Zn^{2+} (Nair 1984b), potassium (Nair et al. 1997), and NH_4-N as well (Mengel 1985).

4.4 Background Information on the Importance of Measuring Zinc Buffer Power

There is a great paucity of published material on the effect of buffer power on availability of heavy metals. Plants obtain most of their fertilizer zinc from reaction products and not applied sources as such, implying that any source of zinc added to soil has to necessarily conform to a chain reaction involving adsorptive, desorptive, and resorptive processes that govern the maintenance of an equilibrium between adequate zinc concentration in the soil solution nearest to the zone of zinc depletion on the one hand and plant uptake on the other. The zinc buffer power defines this. Because the zinc concentration in soil solution is normally very low, the supply to plant roots by mass flow can only account for a very small fraction of plant demand. For instance, with a transpiration coefficient of 300 l kg^{-1} dry matter and a corresponding zinc concentration of 10^{-7} M in the soil solution, approximately 2 mg of zinc can be supplied by mass flow against a demand of 10–30 mg kg^{-1} zinc dry weight of plant tissue.

In calcareous soils, as the zinc concentration is of a much lower order of approximately 10^{-8} M, the supply by mass flow could be very much lower (Marschner 1994), indicating that mass flow can only contribute very negligibly to meet plant needs for zinc. Hence, zinc movement to the plant root surface is principally by diffusion and is essentially confined to a zone around the plant root that hardly extends beyond the root hair cylinder (Marschner 1994). In a review on the mechanism of zinc uptake, Marschner (1994) indicated that flow culture experiments with various species showed adequate ranges of zinc concentration in the range of $6 \times 10^{-8} \times 10^{-6}$ M, which are concentrations greater than those that would be expected in the solution of most soils. He further pointed out that although work using chelate-buffered solutions has indicated adequate zinc concentrations between 10^{-10} and 10^{-11} M, extremely low adequate zinc concentrations required a concomitant excess of about 100 μM zinc chelate as buffer at the plasma membrane of the root cells. This implies a need for an unlimited zinc pool for replenishment of Zn^{2+} at the plasma membrane. When plants grow in soil, it is impossible to expect a zinc buffer of this size to exist, and free Zn^{2+} and chelated-zinc concentrations will be at least threefold lower. Hence, critical deficiency or sufficiency concentrations obtained through research employing chelate-buffered solutions cannot be applied to soil-grown plants.

Most of the work on zinc availability to plants is based on chemical extractions, among which DTPA extraction is the most frequently used. The DTPA extraction quantifies a labile fraction of soil zinc comprising water soluble, exchangeable, adsorbed, chelated, and some occluded zinc. The critical soil level of DTPA-extractable zinc can vary from 0.3 to 1.4 mg kg^{-1} soil, which equates to about 900–4200 g ha^{-1} of zinc in heavy soils and about 600–2800 g ha^{-1} of zinc in light soils in the plough layer (0–20 cm). The crop requirements, on the other hand, are quite small, in the range of 100–300 g ha^{-1} for a total dry matter production of about 10 tons ha^{-1} (Marschner 1994). The inadequacy of DTPA extraction to reflect plant zinc demand shows that other important factors, such as replenishment of soil solution zinc (Nair 1984b), mobility, and transport to the root surface (Wilkinson et al. 1968; Nair et al. 1984), and also the activity of the roots themselves (Wilkinson et al. 1968; Marschner 1994) are involved. Because the zinc buffer power is intricately involved in all three factors, the focus of this review is mainly on that attribute.

As early as three decades ago, it was suggested that colloidal zinc was released by some specific process associated with root activity (Wilkinson et al. 1968). Conditions in the rhizosphere and particularly root-induced changes markedly affect zinc availability. A difference in rhizosphere pH of as much as 2 (higher or lower compared to bulk soil) can be expected to occur as a result of imbalance in ionic uptake. For instance, any acidifying fertilizer such as $(NH_4)_2SO_4$ can result in a net excretion of H+ ions, and others, such as NH_4NO_3, can result in a net excretion of HCO_3^- or OH- ions. Additionally, secretion of organic acids and enhanced CO_2

production will affect rhizosphere pH, and all of the previously mentioned changes will markedly affect zinc availability. However, the scope of this review is confined to the kinetic/dynamic aspects of the changes occurring in the rooting zone mirrored in the zinc buffer power rather than changes in soil reaction in the rhizosphere per se on zinc availability.

The distribution of zinc between the solid and solution phases can be described by the buffer power. The availability of soil zinc to the plant depends on the initial zinc concentration, zinc buffer power, and effective diffusion coefficient (Barber 1984). The Langmuir equation gives the relation between B and C_1 as

$$\frac{C_1}{(x/m)} = \frac{1}{aB} + \frac{C_1}{B}$$

where C_1 is the zinc concentration in the soil solution, x/m is the amount of zinc adsorbed per unit of soil, B is the adsorption maximum, and a is a constant related to the soil's bonding energy for zinc. A straight line is obtained when $C_1/(x/m)$ is plotted against C_1 with a slope of $1/B$ and intercept of $1/aB$. The inverse of $C_1/(x/m)$ is b, the zinc buffer power, where C_1 and x/m are both expressed in volume units (Barber 1984). Using this approach, Shuman (1975) estimated the buffer power values varying from 5 to 100 for four soils representing different major physiographic regions of Georgia. Based on the diffusion model of Drew et al. (1969), Nair (1984a, b) has argued that the C in the equation $U = 2\pi\alpha a C\text{-}\ t$—where, U is the quantity of zinc absorbed per centimeter root length, a is the root radius in cm, α is the root-absorbing power, C- is the average zinc concentration on the root surface, and t is the duration of the absorption period—in fact represents an indirect measure of the zinc buffer power. As we already know, the bulk of zinc uptake is by diffusion (Barber 1984; Elgawhary et al. 1970; Wilkinson et al. 1968). This diffusive process will maintain a concentration gradient in the root zone. This concentration gradient will directly affect zinc uptake because of its effect on the average zinc concentration on the root surface. The zinc buffer power will affect this concentration gradient, because the rate of zinc depletion and/or replenishment is mirrored by it. In a sense, the effective diffusion coefficient and the buffer power are analogous to each other for nutrients that are principally absorbed by the plant root through the diffusive process (Nair 1989). Hence, the crucial question to examine would be the role of zinc buffer power in influencing zinc availability for plant uptake.

4.5 Quantifying the Zinc Buffer Power of Pepper-Growing Soils

Nair (1984b) has used electroultrafiltration to quantify zinc intensity in measuring the zinc buffer power of European soils. Because the procedure is very highly sophisticated and due to its nonavailability in most of the developing countries, a

Table 4.2 Zinc buffer power of pepper-growing soils

Soil	*r* value	*b* value
Peruvannamuzhi	0.8337[a]	0.7824
Thamarassery	0.9304[a]	1.5786
Ambalavayal	0.9604[a]	3.0358

[a]Significant at a confidence level of 0.1%; *b* values represent the zinc buffer power

Table 4.3 Correlation coefficients (*r*) for the interrelationship between routine DTPA soil test versus zinc concentration, zinc uptake, and dry matter production without (1) and with (2) zinc buffer power integration in pepper

Details	1	2
Zinc concentration vs. DTPA test	0.88[a]	0.924[a]
Zinc uptake vs. DTPA test	0.782	0.862[a]
Dry matter production vs. DTPA test	−0.745	0.777[a]

[a]Significant at a confidence level of 0.1%

simple adsorption-desorption equilibrium technique was developed by the author (Nair 2002) to quantify the zinc buffer power of pepper-growing soils. For this, 200 g of representative soil samples from three locations in the state of Kerala, where pepper is extensively grown, were incubated at 60% of the maximum water-holding capacity with graded rates of zinc over a fortnight, by maintaining the water regime to constancy. At the end of the period, the soil samples were extracted with 0.01 M $CaCl_2$ over a 24-h period and the extract tested for zinc using atomic absorption spectrometry. This represented zinc intensity. Separate extractions were made with DTPA, which represented zinc quantity. Data in Table 4.2 indicate that the zinc buffer power values of the experimental soils varied widely.

Data in Table 4.2 bring to light two very important facts. First, the very highly significant correlation coefficients prove that the technique employed allows a precise determination of the zinc buffer power of the soils. Second, the soils varied widely in their zinc buffer power. It is this fact that has to be critically viewed against the existing zinc fertilizer recommendation, emanating from routine soil tests using the universally employed DTPA extraction, to understand the cruciality of the zinc buffer power for precise formulations of zinc fertilizer applications for pepper.

Data in Table 4.3 clearly indicate that integration of the zinc buffer power has improved the relationship between DTPA routine test versus both zinc concentration and zinc uptake. More remarkably, the overall relationship between DTPA test and dry matter production not only improved with zinc buffer power integration in the computations but turned from negative to positive. This clearly shows that the commonly employed DTPA test could only be site specific, but on a larger scale, it is the zinc buffer power that determines plant performance. The experiment had only monitored dry matter production in the pepper plant without taking it to berry formation, because it takes about 3–5 years for berry formation. But, the data in Table 4.3 clearly substantiate the cruciality of the buffer power concept.

Table 4.4 Pepper yield from farmers' fields weighted against zinc buffer power

	Yield (kg $vine^{-1}$)		
Region	Targeted	Actual	Deviation (%)
Peruvannamuzhi	0.241	0.401	+66
Thamarassery	0.490	0.487	+0.6

Note: Target weighting was done against the highest yield obtained from the Ambalavayal region. Note the remarkable closeness between targeted and actual yields in Thamarassery region. Peruvannamuzhi soil is an atypical pepper soil

Data in Tables 4.2, 4.3, and 4.4 have to be viewed in conjunction to precisely understand the cruciality of the buffer power concept. Pepper is the economic mainstay of the state of Kerala. Of late, widespread zinc deficiency has been observed in the state, and the onset of the foot rot due to *Phytophthora* fungi is attributed, to a great extent, to this deficiency in the soils. The scientists at the Indian Institute of Spices Research have made a blanket recommendation of soil application of 23 kg $ZnSO_4$ ha^{-1} when the DTPA-extractable zinc is less than 1.6 ppm (parts per million). In terms of monetary input, this would translate to an equivalent of US $25 ha^{-1} in terms of farmer investment. Because the soils are variously zinc buffered (Table 4.2) a single blanket dose, such as the one that is made, is totally unscientific because, where the farmer needs to apply only lesser quantities of zinc, he would end up applying more if he were to go by the routine recommendation based on the DTPA extraction. To elaborate, the soils from the Ambalavayal region would only require 25% of zinc needed in the soils of the Peruvannamuzhi region because the former has a zinc buffer power nearly fourfold more compared to the latter. In the case of the soils from the Thamarassery region, the quantity required would only be 50%. In fact, the scientists at the Institute have made a gross underestimation of the zinc-supplying power of the soils in the state because their results are all based on the soils of the Peruvannamuzhi region alone, where the Institute has its experimental farm. The soils here are obviously zinc impoverished, but that is not the case in the entire state of Kerala. This also shows that, more often than not, recommendations from experimental stations, when extrapolated on a large scale, can run into problems—especially when such recommendations are based on routine soil testing procedures. These results (Nair 2003) substantiate the earlier ones (Nair 2002) on the importance of measuring the zinc buffer, rather than the routine DTPA-extractable zinc alone, to make precise zinc fertilizer recommendations for wheat in central Asia (Turkey), where farmers have been advised to apply as much as 100 kg $ZnSO_4$ ha^{-1} as a blanket recommendation, without obtaining any tangible and consistent response in wheat yield. Hence, the economic importance of the buffer power concept, as against the routine soil testing procedures, needs hardly any underscoring. Nair et al. (1997) have conclusively demonstrated the importance of the buffer power concept in the case of potassium nutrition for another very important perennial crop and spice, cardamom (*Elettaria cardamomum* M.) in the state of Kerala. Both pepper and cardamom are the staple spices of the state.

References

Adams F (1971) Ionic concentrations and activities in soil solutions. Soil Sci Soc Am Proc 35:420–426
Barber SA (1984) Soil nutrient bioavailability: a mechanistic approach. Wiley, New York
Barber SA (1995) Soil nutrient bioavailability: a mechanistic approach. Wiley, New York
Claassen N, Barber SA (1976) Simulation model for nutrient uptake from soil by growing plant root system. Agron J 68:961–964
Claassen N, Hendriks K, Jungk A (1981) Rubidium—Verarmung des wurzelnahen Bodens durch Maispflanzen. Z Pflanzenern Bodenk 144:533–545
Drew MC, Nye PH, Vaidyanathan LV (1969) The supply of nutrient ions by diffusion to plant roots in soil. I. Absorption of potassium by cylindrical roots of onion and leek. Plant Soil 30:252–270
Elgawhary SM, Lindsay WL, Kemper WD (1970) Effects of EDTA on the self-diffusion of zinc in aqueous solution and in soil. Soil Sci Soc Am Proc 34:66–70
Hendriks L, Classen N, Jungk A (1981) Phosphatverarmung des wurzelnahen Bodens und Phosphataufnahme von Mais und Raps. Z Pflanzenern. Bodenk 144:486–499
Keerthisinghe G, Mengel K (1979) Phosphatpufferung verschiedener Böden und ihre Veränderung infolge Phosphatalterung. Mitt Dtsch Bödenk Ges 29:217–230
Kovar JL, Barber SA (1988) Phosphorus supply characteristics of 33 soils as influenced by seven rates of phosphorus addition. Soil Sci Soc Am J 52:160–165
Lewis DG, Quirk JP (1967) Phosphate diffusion in soil and uptake by plants III ^{31}P-movement and uptake by plants as indicated by ^{32}P-autoradiography. Plant Soil 26:445–453
Lu S, Miller MH (1994) Prediction of phosphorus uptake by field-grown maize with Barber-Cushman model. Soil Sci Soc Am J 58:852–857
Marschner H (1994) Micronutr News Inf 14:3–5
Mengel K (1985) Dynamics and availability of major nutrients in soils. Adv Soil Sci 2:65–131
Nair KPP (1984a) Towards a better approach to soil testing based on the buffer power concept. In: Pierre-Martin P (ed) Proceedings of the sixth international colloquium for the optimization of plant nutrition, vol 4, September 2–8. Montpellier, France, pp 1221–1228
Nair KPP (1984b) Zinc buffer power as an important criterion for a dependable assessment of plant uptake. Plant Soil 81:209–215
Nair KPP (1989) Comments on Phosphorus supply characteristics of 33 soils as influenced by seven rates of phosphorus addition. Soil Sci Soc Am J 53:984
Nair KPP (1992) Measuring P buffer power to improve routine soil testing for phosphate. Eur J Agron 1(2):79–84
Nair KPP (2002) Sustaining crop production in the developing world through the Nutrient Buffer Power Concept. In: Proceedings of the 17th World Soil Science Congress, Bangkok, Thailand, August 14–21, vol 2, p 652
Nair KPP (2003) Measuring Zn buffer power to predict Zn availability. J Plant Nutr (in Press)
Nair KPP, Mengel K (1984) Importance of phosphate buffer power for phosphate uptake by rye. Soil Sci Soc Am J 48:92–95
Nair KPP, Ram N, Sharma PK (1984) Quantitative relationship between zinc transport and plant uptake. Plant Soil 81:217–220
Nair KPP, Sadanandan AK, Hamza S, Jose A (1997) The importance of potassium buffer power in the growth and yield of cardamom. J Plant Nutr 20(7 & 8):987–997
Nye PH (1979) Diffusion of ions and uncharged solutes in soils and clays. Adv Agron 31:225–272
Schüller H (1969) Die CAL-Methode, eine neue Methode zur Bestimmung des pflanzen verfügbaren Phosphates in Böden. Z Pflanzenernähr Bodenkd 123:48–63
Selim HM (1992) Modeling the transport and retention of inorganics in soils. Adv Agron 47:331–384
Shuman LM (1975) The effect of soil properties on zinc adsorption by soils. Soil Sci Soc Am Proc 39:454–458

Soil Science Society of America (1965) Glossary of soil science terms. Soil Sci Soc Am Proc 29:330–351, Committee Report

Sparks DL (1987) Potassium dynamics in soils. Adv Soil Sci 6:2–63

Sparks DL (1989) Kinetics of soil chemical processes. Academic, San Diego

Wilkinson HF, Loneragan JF, Quirk JP (1968) The movement of zinc to plant roots. Soil Sci Soc Am Proc 32:831–833

Chapter 5
Establishing a Pepper Plantation

Abstract The chapter describes the establishment of pepper plantations with specific reference to the Indian experience and the Indonesian experience. Both India and Indonesia are the major pepper producers in the world. However, while the latter scores on productivity in pepper production, the former holds the place of honor for quality.

Keywords Plantation · India · Indonesia · Production · Quality · Pepper

5.1 The Indian Experience

From a commercial viewpoint, establishing a good pepper plantation is of great significance. Unlike in countries such as Thailand, Malaysia, and Brazil, where the life span of a pepper plantation is only 10–15 years, at the end of which replanting is done, in India a pepper plantation can last for more than a quarter century. The prime reason for the shorter duration of the pepper plantations in other countries is because pepper is trailed on nonliving standards (support), while in India pepper is always trailed on live support, such as *E. indica*. The starting point to establish a good pepper plantation is the establishment of good planting material. With the exception of India, most other countries use the orthotropic (upward-climbing) shoot as the planting material. This is the best planting material because it results in vigorous plants and bear fruiting laterals right from the base itself and yields a lot earlier. In India, however, the use of orthotropic shoots is greatly restricted because of nonavailability.

In earlier times, pepper planters used either runner shoots or climbing orthotropic shoots to coincide with the onset of the southwest monsoon in India. These days, prerooted cuttings from runner shoots are used instead. When hanging shoots are used as planting material, they result in the formation of weak plants. Quite often, pruned material—pepper—is pruned 5–6 months after planting, and once again, after a year, is also used as planting material. These are pruned stem cuttings.

K. P. Nair, *The Geography of Black Pepper (*Piper nigrum*)*,
https://doi.org/10.1007/978-3-030-52865-2_5

Photo 5.1 Bush Pepper in pot

Because pepper is only propagated through the type of planting material described earlier, there is a huge worldwide demand (running into millions) for ready-to-use planting material. Production of prerooted cuttings in light polyethylene bags, PVC (poly vinyl chloride) bag is the surest way of mass producing material once the demand for planting material escalated in India. Their production is as follows. Runner shoots from high-yielding and healthy plants are kept coiled at the base of wooden pegs fixed at the bottom of the plant so as to prevent the growing shoots from spreading on the soil and striking roots. These runner shoots can be separated from the mother plant in winter (January–February) by snapping them using scissors and are disinfected by dipping them in a fungicide solution, such as oxychloride or a Bordeaux mixture for 1 min; they are later surface dried and cut into three noded strips after clipping the leaves; they are then planted either in raised soil beds or PVC bags filled with a mixture of soil, sand, and some farmyard manure, which will provide enough nutrients in a growth substrate. Application of growth regulators, such as indole butyric acid (IBA), by dipping the cuttings in the solution has been found to enhance root proliferation (Pillai et al. 1982; Suparman and Zaubin 1988). Care must be taken to minimize fungal infection arising from the nursery bed Photo 5.1.

In Sri Lanka, growing planting material on split bamboo led to the production of good planting material on a large scale. Popularly known as the *bamboo technique* (Bavappa and Gurusinghae 1978), the method consists of first digging a trench (60×40 cm^2) of convenient length, filling it with a rooting medium—preferably forest soil, sand, and farmyard manure in a 1:1:1 ratio mixture; split bamboo poles

Fig. 5.1 Split bamboo technique developed in Sri Lanka for vegetative pepper propagation

(1.25–1.5 m long) or PVC pipes, both of about 10 cm in diameter, are then sunk into the trenches at 45° angles using a strong middle support. Rooted cuttings, one each per bamboo, are planted in the trench and allowed to trail onto the bamboo or PVC poles. The lower portion of the pole is filled with a rooting medium (of weathered coir dust and farmyard manure in a 1:1 ratio), and as the vine grows, it is tied to the support. Vines are irrigated daily, and to stimulate rapid growth, a nutrient solution consisting of nitrogen, phosphorus, potassium, and magnesium (1 kg urea, 0.75 kg superphosphate, 0.5 kg muriate of potash [MOP], and 0.25 kg magnesium sulfate all mixed in 250 l of water) can be applied at the rate of 0.25 l per vine every fortnight. It takes 3–4 months for the vine to reach the top of the pole, and when it does, the terminal bud is nipped off, and the stem is crushed about three nodes above the base to activate the axillary buds; after about 10 days, each vine is cut at the crushed point, and this is ready for transplanting into PVC bags, filled with the rooting mixture as detailed earlier. Care should be taken, while planting in PVC bags, not to injure the roots and not let the axil be below the rooting medium. The PVC bags are placed in a cool place and covered with a light PVC sheet (200 gage) to retain high internal humidity. Buds start to develop in about 3 weeks' time, and then the PVC bags can be shifted to a place with shade. Regenerated shoots from the stumps can be trained as before, which permits the production of a continuous stream of rooted cuttings as planting material. Four harvests are possible in 1 year, with a multiplication ratio of 1:40. A prolific root system develops, and field establishment is very good. In Sarawak (Malaysia), on average, 54 rooted cuttings could be obtained (Ghawas and Miswan 1984), and the method has been successfully field tested in the experimental farm of the Indian Institute of Spices Research. Figure 5.1 shows the bamboo technique developed in Sri Lanka for vegetative pepper propagation.

Pepper can be planted on flat surfaces and slopy land. If it is the latter, slopes facing south should be avoided. On slopy land, it is advisable to plant along the contour lines to arrest topsoil loss due to runoff water during monsoon. An important feature of pepper plantations in India is that the vines are trailed on living standards, mainly *E. indica* or *E. lithosperma* (a thorny tree and the trailing vines have strong footholds on the thorns), *Gliricidia*, *G. robusta* (Silky Oak), *G. pinnata*, and *Ailanthus* spp. In homestead plantations, coconut and areca nut trees act as the main crop in addition to being the standards. In the southern state of Karnataka, pepper is grown along with coffee, where coffee is the main crop and the forest trees act as the standards. *Gliricidia* is the most common standard in Sri Lanka and Malaysia. Pepper planting should precede the onset of monsoon. Two or three three-noded rooted cuttings, prepared as explained earlier, are planted in shallow pits on the northern side of the standard. If the cuttings are unrooted, four to five could be planted. At least one node should be below the soil for the establishment of roots. To accelerate growth, pits can be filled with topsoil, farmyard manure, rock phosphate, and neem cake mixture (Pillai 1992). Plant density is around 1970–2000 ha^{-1} (George 1982) and can go up to 5000 ha^{-1} when nutrient and carbon stress can be observed (Reddy et al. 1992).

In order to support the climbing vine on the standard, it should be tied as frequently as necessary using jute thread. Pruning of the terminal part, though not extensively practiced in India, encourages growth and bearing (Kurien and Nair 1988). When a pepper plantation is established, starting from the beginning of the year, specific operations are to be followed. In the first 2 months, starting from January, when harvesting is completed, pruning the hanging shoots, tying vines, and mulching the basin are done. In the following 2 months, diseased vines are removed, and the pits could be filled with topsoil, 1 kg of lime, and 5 kg of farmyard manure mixture. Southwest monsoon begins in May and June, and with this a mixture of 5 kg of farmyard manure, 250 g of rock phosphate and 1 kg of neem cake should be applied. In June and July, a Bordeaux mixture (1%) is applied, and drenching the pits with copper oxychloride (0.2%) is done. The damaging pollu beetle (*Longitarsus nigripennis*), a deadly insect that leaves the berries hollow (in the regional language, *Malayalam*, of Kerala State, "pollu" means hollow), can be controlled by spraying endosulfan (1.5 ml l^{-1}). Application of 50 g of urea and 120 g of MOP should also be done. In September and October, 50 g of urea, 120 g of MOP and 50 g of $MgSO_4$ are applied. Additionally, a Bordeaux mixture spray (1%) is also applied. Tying of vines should be continued. In November and December, phytosanitary measures, such as removal of infected plants, drenching with oxychloride (0.2%), mulching, and shading the young vines, must be done.

It is uncommon to irrigate a pepper crop, except in Thailand where it is grown as an irrigated crop. Nevertheless, pepper plants can seriously suffer due to moisture stress in the summer months, especially in India, when daytime temperatures in April and May can soar to 35 °C or higher. A field study conducted at the Pepper Research Station in Panniyur, Kerala State, over an 8-year period (1988–1996), indicated that irrigation at an IW/CPE ratio of 0.25 increased pepper yield as much as 90%, the positive effect being manifested most in Panniyur-1 and Karimunda (Cultivar), both released from this station (Satheesan et al. 1997).

An interesting aspect of pepper cultivation is the possibility of growing other crops, in addition to pepper, in the same field or plot of land. This practice is more in common when it is the case of homestead farming, where income from the accompanying crop adds to the profit from the main pepper crop. However, this is not a general practice on large pepper plantations because raising an additional crop makes demands of its own, in terms of nutrient input, pest management, and so forth. It is important to note that companion crops to pepper must best be raised in the prebearing stage of pepper, which at best can extend from 3 to 5 years (at the most). This is because, when fully grown, pepper canopy could completely shade the accompanying crop, and the latter would hardly yield. An experiment at the Indian Institute of Spices Research (Sadanandan 1994) reported that an accompanying crop of banana (cv. *Mysore poovan*) fetched an additional income of approximately US \$460 ha^{-1}. Pillai et al. (1987) have obtained comparable results.

The recent glut in the pepper market and slump in prices has prompted many farmers to look to companion crops to generate additional income. Simultaneously, the falling prices of coconut have prompted coconut farmers to use the coconut trees as a live standard to raise pepper. In this connection, an interesting experiment conducted at the Central Plantation Crops Research Institute (CPCRI) at Kasaragod, Kerala State, which has a mandate for research—primarily with regard to coconut—merits mention.

Pepper vines (Panniyur-1) were trailed on 60-year-old coconut palms, and when the vines reached a height of 4–5 m, further growth was restricted to enable farmers from climbing the palms to pluck the nuts. On average, 2 kg of dry pepper vine were obtained. Companion cropping with areca nut has also been tried. An areca nut-pepper combination gave 3832 kg ha^{-1} of dry dehusked areca nut and 1418 kg ha^{-1} of dry pepper (Nair and Gopalasundaram 1993) from 1000 vines ha^{-1}. Companion cropping is common in Brazil as well, where an array of companion crops—such as rubber, cocoa, orange, lemon, and clove—are extensively tried. Though growing mixed companion crops in pepper gardens is a practice in many situations, it needs to be clearly understood that when interspecific crops are grown in association, the biological implications could be diverse. This diversity might originate not only in nutrient absorption patterns, moisture stress relationships, and the like but can also extend to other areas, such as susceptibility/resistance to specific pests and diseases. These are the areas that deserve thorough scientific scrutiny before value judgment can be made on the suitability or otherwise of growing companion crops with the main crop of pepper. Figure 5.2 shows a field view of an Indian pepper plantation.

5.2 The Indonesian Experience

Within Asia, next to India, Indonesia is the major pepper grower. Pepper was the first spice Indonesia traded with Europe through Persia and Arabia and prior to World War II, 80% of world production was controlled by Indonesia. It was the Japanese

Fig. 5.2 A field view of an Indian pepper plantation

occupation during the war that left many pepper plantations uncared for, with the resulting decline in production. Compared to the prewar period when the country could boast of more than 20 million vines, after the war it came down to just about a lakh (one hundred thousand) of vines. The most interesting characteristic of pepper cultivation in Indonesia is that most of the pepper gardens are owned by small farmers, unlike in India where large plantations can be found. Currently, the main producing areas are Lampung and Bangka, with Lampung in the lead. During the last quarter of the past century, there was a perceptible decline in cultivated area in Lampung, with production shifting to Bangka, east and west Kalimantan, and Sulawesi. In Bangka, pepper was first grown in Mentok; Mentok white pepper had a good world market.

In Indonesia, pepper is grown in various types of soil, such as andosol, grumsol, latosol, podsol, and regusol. Well-drained alluvium rich in humus, pH above 5.8, is ideal for pepper (Zaubin and Robbert 1979). In general, pepper soils in Indonesia are poor in fertility, texture, and structure and low in organic matter. The two main pepper-growing areas—Bangka and west Kalimantan—have poor, reddish brown, and sandy soil. Soils in Lampung are reddish brown latosols and contain more organic matter than the soils of Bangka. In Indonesia, the planting material is made up of cuttings from orthotropic shoots. Normally, cuttings with five to seven internodes are used, with two to four nodes buried in the soil. Cuttings are planted at an angle of 35°–45° with reference to the standard. For the first 3–4 weeks, the cuttings must be shaded from the sun. Cuttings are planted in pits $0.6 \times 0.8 \times 0.6$ m in size. Rooted cuttings are also used. Precautions explained earlier must also be adhered to in order to protect the cuttings. The procedures used in the case of prerooted cuttings, except that they have a better ability to withstand heat exposure from sunshine, are the same as described earlier. Such planting materials are selected from the nursery, and care should be taken that only well-rooted and well-grown cuttings in PVC bags are transplanted in the main field.

Unlike in India, Sri Lanka, and the Philippines, in Indonesia the types of standards used are both living and nonliving. In the former category, we have *Gliricidia*, *Dadap*, and *Kapok*, while in the latter category, we have hard timber poles (made of iron or wood) or concrete poles. Experience shows that concrete poles result in poor growth and productivity. In Lampung, small farmers use living standards, while nonliving ones are used in Bangka and east and west Kalimantan. There is a gradual shift to living standards because nonliving ones, of late, have become more expensive.

Pepper plantations are usually owned by small farmers, with holdings ranging between 0.2 and 1.5 ha. On average, the size is 0.65 ha. In general, pepper cultivation is intensive in Bangka and south Sumatra, while in Lampung it is extensive, where pepper is planted under live standards and input rates, in terms of fertilizer, pest management, and the like, are meager.

The Indonesian pepper industry is mostly owned by small farmers, and planting pepper on a plantation scale did not last long in Java. The nature of pepper cultivation and the degree of involvement of labor and capital have played a very important role in determining the mode of the pepper farming system in Indonesia. However, it is a tribute to the pepper industry in Indonesia that the IPC office is headquartered in Jakarta. Figure 5.3 shows a field view of an Indonesian pepper plantation.

Fig. 5.3 A field view of an Indonesian pepper plantation

References

Bavappa KVA, Gurusinghae (1978) Rapid multiplication of black pepper for commercial planting. J Plantation Crops 6:92–95

George CK (1982) Pepper cultivation in Malaysia. Indian Cocoa Arecanut Spices J 5:75–76

Ghawas M, Miswan J (1984) Satu Kaedah memprecepatkan Pembiakn lada hitam (*Piper nigrum* L.). Teknol Pertanian MARDI Res J 5:50–54

Kurien S, Nair PCS (1988) Effect of pruning on yield in pepper (*Piper nigrum* L.). Agric Res J Kerala 26:137–139

Nair MK, Gopalasundaram P (1993) Nutrient management in tree based cropping system. In: Singh M, Mishra MK (eds) Potassium for plantation crops. Potash Research Institute of India, Gurgaon, pp 200–211

Pillai VS (1992) Management of improved cultivars of black pepper. In: Sharma YR, Devasahayam S, Anandaraj M (eds) Black Pepper and Cardamom: Problems and Prospects. Indian Society of Spices, Calicut, pp 17–19

Pillai VS, Ali AB, Chandy KC (1982) Effect of 3-IBA on root initiation and development in stem cuttings of pepper (*Piper nigrum* L.). Indian Cocoa Arecanut Spices J 6(1):7–9

Pillai VS, Sasikumaran S, Nambiar PKV (1987) N, P and K requirement of black pepper. Agric Res J Kerala 25(1):74–80

Reddy BN, Sadanandan AK, Sivaraman K, Abraham J (1992) Effects of planting density on black pepper on yield and nutrient availability in soil. J Plant Crops 20(Suppl):10–13

Sadanandan AK (1994) Nutrition of black pepper. In: Chadha KL, Rethinam P (eds) Advances in horticulture. Vol 9: plantation and spice crops (Part I). Malhotra, New Delhi, pp 423–456

Satheesan KN, Rajagopalan A, Sukumarapillai V, Jagadeesh Kumar TN, Mammootty KP, Unnikrishnan Nair PK (1997) Influence of irrigation levels on yield, quality and incidence of pests and diseases in black pepper (*Piper nigrum* L.) (Abs.). In: National seminar on water management for sustainable production and quality of spices. Madikheri, Karnataka State, India, p 21

Suparman U, Zaubin R (1988) Effect of defoliation, IBA and Saccharose on root growth of black pepper cuttings. Ind Crop Res J 1:54–58

Zaubin R, Robbert (1979) Pengaruh keasaman tanah terhadap pertumbuhan tanaman lada. Pebr Littri 33:1987

Chapter 6
Pepper Pests and Their Control

Abstract The chapter specifically describes several pepper pests, both bacterial and fungal, including virus and mycoplasma, besides insects and their chemical control. The pepper plant is attacked by diseases caused by fungi, bacteria, virus and mycoplasma, insects, and nematodes. Among the diseases, the sudden wilting of the pepper plant caused by *Phytophthora* is the most deadly. The fungus causing this dreaded disease is *Phytophthora capsici*. Both chemical control and organic amendments control these infestations. A recent development in the control of *Phytophthora* is the employment of biological organisms, primarily from a nonfungal background. Mention must be made to vesicular arbuscular mycorrhizae.

Keywords Phytophthora · Nematode · "Pollu beetle" · Vesicular arbuscular mycorrhizae (VAM)

The pepper plant is attacked by several diseases, caused by fungi, bacteria, virus and mycoplasma, insects, and nematodes. Nutritional disorders aggravate the impact of the pests—be they caused by any one of the preceding disorders or soil and/or environmentally triggered. However, it is important to make a distinction between, for example, a deadly disease that is caused by a fungus, such as *Phytophthora* spp. (from which specific symptoms in the plant will originate), and a nutrient deficiency in the soil that may lead to an imbalance of the specific nutrient in the plant and that will aggravate the symptoms of the disease subsequently. It is important to recognize that of all the diseases a pepper plant may suffer from caused by the *Phytophthora* fungus are the most devastating. Hence, this review will primarily focus on the foot rot disease caused by *Phytophthora*, which is the major disease in India, Malaysia, and Indonesia causing serious damage to pepper production (Holliday and Mowat 1963; Kueh and Sim 1992b; Manohara et al. 1992; Sarma et al. 1992). In India, although wilt disease was reported as the major cause of pepper crop loss, *Phytophthora*, as its causal agent, was first reported only in 1966 by Samraj and Jose (1966). There are, in fact, 17 diseases that cause loss in yield, but among these *Phytophthora* foot rot and slow decline—which were originally referred to as quick

K. P. Nair, *The Geography of Black Pepper (*Piper nigrum*)*,
https://doi.org/10.1007/978-3-030-52865-2_6

Photo 6.1 Pepper harvesting by men labourers (harvested berries in back pack)

wilt and slow wilt, respectively—cause the most damage (Das and Cheeran 1986; Nambiar 1978; Nambiar and Sarma 1977). *Phytophthora capsici* is the causative fungus. The infection occurs aerially as well as through soil, and the severity of infection depends on the site of infection (Anandaraj and Sarma 1995; Mamootty 1978). When it is a foliar infection, one to many dark spots appear (with characteristic fimbriate margins), which advance and later coalesce, leading to defoliation even before the lesions spread to the entire lamina. The base of the vines can also be infected. On tender shoots, the fungus profusely sporulates forming a white covering, and when the infection reaches the stem, abrupt wilting of the entire plant takes place. Spikes, when infected, lead to the formation of blackened berries. When the infection is below the soil on roots, rotting and degeneration start, leading to yellowing, defoliation, and drying up of the plants. Feeder root infection reaches the collar through main roots, which causes the characteristic foot rot, hence the name (Anandaraj et al. 1994) Photo 6.1.

Because foot rot results in sudden wilting followed by death of the plant, the disease was earlier referred to as wilt or quick wilt and the disease-causing fungus was identified as *Phytophthora palmivora* var *piperis* by Samraj and Jose (1966) and as *P. palmivora* by Holliday and Mowat (1963). In India, though pepper is traditionally grown in the Western Ghats in the states of Kerala, Karnataka, and Tamil Nadu, it has been introduced to nontraditional pepper-growing areas, such as the Andhra state and also the northeastern states, and the incidence of foot rot has been reported from these places as well (Sarkar et al. 1985). Up to 30% crop loss has been reported (Nambiar and Sarma 1977; Samraj and Jose 1966). The onset and spread of foot rot have been found to be closely correlated to a wet and cloudy atmosphere prevalent during the southwest monsoon period in the state of Kerala. A combination of factors, such as daily rainfall of 1.6–2.3 cm, ambient temperature ranging from 22.7 to 29.6 °C, relative humidity of 81–99%, and daily sunshine duration of 2.8–3.5 h favors the spread of the disease (Ramachandran et al. 1988b, 1990).

Inasmuch as infection through soil substrate is concerned, the prime source of infection is contaminated soil where new planting is done. The inoculum can survive in the soil for up to 19 months in the absence of a host plant (Kueh and Khew 1982). The pathogen is normally concentrated to a depth of 30 cm (Ramachandran et al. 1986), and the severity of concentration diminishes as one moves deeper and away from the infected vines. Management of foot rot is, perhaps, the most challenging aspect of pepper cultivation. There are various aspects to the control measures, such as cultural, phytosanitary, biological, and chemical. The most important among the cultural measures refer to maintenance of good drainage. Stagnant water leads to anaerobic conditions triggering germination of pathogen propagules. A decrease in the production of phenol oxidase, phytoalexin, and fixed nitrogen, combined with suppressed mycorrhiza, leads to host plant susceptibility in anaerobic conditions (Drew and Lynch 1980). Good drainage blocks the buildup of *P. capsici*. Phytosanitation assumes an important role in controlling the disease. Because the infected plants serve as the foci of infection (Zadocks and Van den Bosch 1994), their removal is most crucial in controlling the disease. The initial occurrence and spread of the disease is nonrandom and tends to cluster around the previously infected plants (Anandaraj 1997); this is the reason such infected plants have to be removed forthwith. Further, as pepper grows on living standards, they develop a canopy of their own, generating a microclimate different from that of the ambient situation, with high humidity and low temperature, which is ideal for *P. capsici* to multiply fast and infect.

Branches of the living standards should be regularly lopped, especially during the rainy season, which facilitates the penetration of direct sunlight that leads to heating the soil and thereby reducing the humid surroundings, thus acting as a check for the multiplication of the pathogen. The loppings could be used as surface mulch that, while adding to the organic matter of the substrate soil, can also check weed growth. Because the prime source of the initial inoculum of *P. capsici* in pepper plantations is the contaminated soil, infection of the foliage can occur through soil splashed onto tender shoots trailing on the ground and from these shoots onto other parts of the vines through rain splashes (Ramachandran et al. 1990). In order to arrest soil splashes, legume or grass cover mulch is ideal (Ramachandran et al. 1991; Sarma et al. 1992). However, studies in Malaysia (Ahmed 1993) indicate that pepper grows better in clean-weeded plots, unlike in those with a cover crop of *Desmodium trifolium*. The Indian experience shows (Anandaraj 1997) that after the cessation of the southwest monsoon, it would be ideal to clean up the pepper plots of the existing weeds and rake up the soil, which was seen to check the spread of the pathogen.

Chemical control of *Phytophthora* has revolved around the use of systemic fungicides, such as metalaxyl formulations like Ridomil granules or Ridomilziram (Kueh and Sim 1992a; Kueh et al. 1993; Ramachandran 1990; Ramachandran and Sarma 1985; Ramachandran et al. 1988a, 1991; Sarma et al. 1992) and/or the classical Bordeaux 1% spray mixture. Pasting the collar with Bordeaux mixture and drenching the basin/trench around the vine with oxychloride have also been effective (Lokesh and Gangadarappa 1995; Malebennur et al. 1991; Nair and Sasikumaran 1991; Ramachandran et al. 1991). Owing to its high cost, however, chemical control is not very popular with small farmers. Further, Coffey (1991) has expressed doubts about the efficacy of chemical control of the soilborne diseases caused by *Phytophthora*.

In addition to the preceding measures of control, there is also the possibility of the use of organic amendments and biological control. Inasmuch as the addition of organic amendments to the soil are used to control the *Phytophthora* spp., the objective is to raise the growth of saprophytes (Kueh and Sim 1992a), and it has been observed that the enhanced activity of saprophytes checks the growth of *P. capsici* dramatically (Anandaraj 1997). Oil cakes—such as those obtained after expelling oil from neem seed, groundnut, and coconut—can beneficially be used to control the growth of the disease-causing fungus (Nair et al. 1993; Sadanandan et al. 1992). In addition, even chicken manure has been found to be useful. Applications of organic amendments such as the ones cited earlier will have a twofold beneficial effect: (1) enhancing the nutrition of the vines and (2) encouraging the growth of saprophytes that check the population of *P. capsici*. Both complement each other in the overall control of the fungal growth.

A recent development in the control of *Phytophthora* is the employment of biological organisms, primarily from a fungal background. Mention must be made of vesicular arbuscular mycorrhizae (VAM), *Trichoderma*, and *Gliocladium* (Anandaraj and Peter 1996; Anandaraj and Sarma 1994b; Datta 1984; Manjunath and Bagyaraj 1982; Ramesh 1982). Soilborne pathogens are amenable to biological control (Cook and Baker 1983). Because *P. capsici* is a soilborne fungus and the prime source of contamination is infected soil, growth of antagonistic fungi would check the buildup of the pathogenic fungus. Interesting work on isolation and mass multiplication of inexpensive carrier media for field application of microorganisms having biological control properties has been carried out (Anandaraj and Sarma 1995; Sarma et al. 1996). An inexpensive carrier medium for the culture of these beneficial organisms is water contained in mature coconuts that is thrown away after shelling. Both *Trichoderma* and *Gliocladium* have been found to grow well in this medium (Anandaraj and Sarma 1997).

Interest in VAM has been mainly focused on its ability to solubilize native phosphorus. Its use in the control of pepper diseases was started when it was first observed that VAM suppressed the foot rot caused by *Phytophthora* in citrus (Davis and Menge 1981; Davis et al. 1978). Soil incorporation of VAM can also be done along with *Azotobacter* and *Azospirillum*, which, besides controlling the pathogen, have beneficial growth effects (Bopaiah and Khader 1989; Govindan and Chandy 1985). The positive effect of VAM is through the alteration of the nature of root exudates and the rhizosphere microflora—which is generally termed the "mycorrhizosphere effect"—and there is extensive research to substantiate this view (Dehne 1982; Graham 1982, 1988; Linderman 1988). The positive effects of VAM have not only enhanced plant growth but have also suppressed disease-causing pathogens, as reported by Ewald (1991) and Graham and Egel (1988). In view of the overall positive effect of VAM, it has been recommended that VAM be incorporated into the soil right from the nursery stage and onward so that the beneficial effects of enhancing growth and suppression of pathogenic effects are obtained from the early stages (Sarma et al. 1996). Field experience shows that strict phytosanitation in the nursery is a very important prerequisite in subsequently installing a disease-free pepper plantation.

Next in importance to diseases caused by pathogenic fungi are the diseases caused by nematodes. Crop productivity is seriously hampered by plant parasitic nematodes, and pepper is no exception. Of the 15,000 nematode species described so far, 2200 are plant parasitic. Quite often, nematode damage goes unnoticed. Because nematode infestation starts at the root surface, pepper vines so infected will subsequently suffer from a number of secondary complications, and quite often these symptoms can be mistaken for nutritional or other physiological disorders. When susceptible crops are continuously grown (in the following year), they become a good breeding ground for disease-causing nematodes. A compilation of plant parasitic nematodes associated with pepper in the major pepper-growing countries listed 48 species that belong to 29 genera (Sundararaju et al. 1979), while 54 species that belong to 30 genera were listed by Ramana and Eapen (1998). In India, 17 genera of nematodes associated with pepper were listed from the two primary pepper-growing states of Kerala and Karnataka (Sundararaju et al. 1980), while in Indonesia, 14 genera were listed (Bridge 1978; Mustika and Zainuddin 1978). Among the nematode species, *Meloidogyne incognita* and *Radopholus similis* are the most important and cause the most severe damage. The former is commonly known as the "root knot nematode," while the latter as the "burrowing nematode." On a global scale, the former is the most devastating.

The root knot nematodes have a specialized and complex relationship with the host plant, and they are sedentary obligate endoparasites. Their infestation of the vine results in the formation of elongated swellings on the thick primary roots due to multiple infections and typical knots or galls on either secondary or fibrous roots because of hypertrophy and hyperplasia of the infected tissues. The name *root knot* is derived because of these typical knots on the roots. In thick primary roots, a number of adult females with egg masses localize deep below the epidermis, and the entire length of the root turns into a gall and appears smooth with infrequent swellings

(Mohandas and Ramana 1987). Because nematodes feed on vascular tissues, a disruption in the arrangement and continuity of the vascular bundles leads to impaired movement of both water and nutrients, and as a consequence, the plant is vitally affected. When infestation is severe, a large amount of root mass is lost due to eventual decay of the galled roots, which in turn very adversely affect the entire vine (Mohandas and Ramana 1991; Siti Hajijah 1993). When root knot infestation takes place, yellowing of the foliage occurs, resulting in stunted growth and eventual decline of the vine. Dense yellowing of interveinal areas—with the deep green veins prominently visible (Ramana 1992)—can often be mistaken for nutritional deficiency. When vines are infested with *M. incognita*, certain impaired physiological reactions, such as lowered absorption and translocation of phosphorus, potassium, zinc, magnesium, copper, calcium, and manganese and their accumulation in leaves, have been observed (Ferraz et al. 1988) as well as a reduction in total chlorophyll content of leaves (Ferraz et al. 1989). These changes lead to stunted growth of the vines. When vines were inoculated with *M. incognita* inoculum, a high concentration of total phenols without expression of any resistance to the pest was observed (Ferraz et al. 1984). Additionally, several changes in amino acids, organic acids, and sugars were also observed in vines infested with the nematode (Freire and Bridge 1985a).

The burrowing nematode (*R. similis*) is an obligate and migratory endoparasite, is extensively found in both tropical and subtropical regions of the world, is a serious pest of pepper (Holdeman 1986), and has a wild range of hosts (about 370 plant species). The existence of *R. similis* was first reported on the banana host by Nair et al. (1966) in Kerala State. The nematode feeds on cortical tissues and produces elongated dark brown necrotic lesions on the roots at the site of infection. The nematode pushes through the cell wall of each cell after draining its contents, and this burrowing phenomenon results in the formation of tunnels in the root tissues. The nematode derives its name from this trait of burrowing. When infestation is severe, many lesions coalesce and encircle the root cortex; because of this damage to the cortical cells, the root portion distal to these lesions gradually disintegrates. The vines tend to produce new roots, which in turn get infected, leading to the formation of a bunch of decayed root mass (Mohandas and Ramana 1987). Yellowing of leaves, defoliation, and generally stunted growth result from infestation.

Another important disease, popularly known as the "slow decline disease," caused by a combination of the two nematodes *M. incognita* and *R. similis* and the fungus *Fusarium* spp., has been the cause for widespread destruction of pepper in several countries. The disease is also known as "yellows." It was the cause of major pepper devastation in the Bangka Islands of Indonesia in the 1950s (Christie 1959) and also in Guayana (Biessar 1969), India (Nambiar and Sarma 1977), Malaysia (Kueh 1979, 1990), Brazil (Ichinohe 1975; Sharma and Loof 1974), and Thailand (Bridge 1978; Sher et al. 1969). The disease was first observed on the islands of Bangka (Van der Vecht 1950) in Indonesia and later spread to the islands of Belantung. The disease caused losses of up to 30% annually (Sitepu and Kasim 1991). The disease is widely prevalent in Sarawak (Malaysia). The total life span of the vines can be reduced to 8–10 years, compared with the normal 2530 years

(Varughese and Anuar 1992). In Cambodia, where pepper is predominantly found in the Kampot region, the disease was responsible for bringing down vine population from 2.5 million in 1942 down to 0.5 million in 1953 (Hubert 1957). In the predominantly pepper-growing state of Kerala in India, there is no precise estimation of crop loss, but Menon (1949) reported a 10% loss of vines. It has been observed that plants infested by root knot nematodes are more susceptible to the fungus *Phytophthora* (Winoto 1972).

Foolproof management of nematode infestation is an elusive target. It is important to recognize that the primary source of infestation is nurseries. Solarization and fumigation are effective in controlling nematode infestation in nurseries. Additionally, incorporation of biocontrol agents, such as VAM, into solarized soil has been found to be an encouraging method for the control of nematode infestation (Anandaraj and Sarma 1994a, b; Sarma et al. 1996). There are different angles to the control regime. For instance, in plantations it has been observed that chemicals released during decomposition of organic manure—for instance, Azadirachtin from neem cake and Ricin from castor cake—are toxic to nematodes (Stirling 1991). Mulching has been found to have a positive effect on nematode control (Ichinohe 1980, 1985; Wahid 1976). Hubert (1957) obtained very encouraging results in nematode control using *Eupatorium* mulch. Of the various aspects of developing resistance/tolerance to nematodes, the most reliable is the development of plant resistance. But, to date, there are no varieties or cultivars that are totally resistant to nematode attack. Obviously, the focus of pepper breeding has not been on developing absolute nematode-resistant varieties. In fact, polygenic-horizontal resistance is more important than monogenic-vertical resistance (Fassuliotis and Bhatt 1982). Existence of physiological races or pathotypes in nematodes, in particular *Meloidogyne* spp., is another factor that must be keenly focused on in any breeding program in pepper for nematode resistance. Currently, efforts focus on developing tolerance to nematode infestation rather than developing absolute resistance. Also, efforts are being made to screen the currently available cultivars for their reaction to nematodes. For instance, in Sarawak (Malaysia), Kueh (1986) found cv. *Uthirancotta* to be the most susceptible cultivar, while cultivars such as *Balancotta*, *Belantung*, *Cheriaka-niakkadan*, *Jambi*, and *Kalluvally* are less susceptible to root knot nematodes under field conditions. Mustika (1990, 1991) found cv. *Kuching* to be tolerant to *M. incognita* and *R. similis*, compared with *Kalluvally*, *Jambi*, and *Cunuk*. In India, none of the cultivars tested were found to be resistant to *M. incognita* and *R. similis* compared with *Kalluvally*, *Jambi*, and *Cunuk*. In India, none of the cultivars tested were found to be resistant to *M. incognita* (Jacob and Kuriyan 1979; Koshy and Sunderaraju 1979; Ramana and Mohandas 1986) or *R. similis* (Ramana et al. 1987; Venkitesan and Setty 1978). However, it has been found that among the Indian cultivars, *Pournami*, which is a selection from the germplasm collection at the Indian Institute of Spices Research, is tolerant to *M. incognita*. This has been supported by field evidence that shows that where this cultivar is cultivated, only a smaller population of nematodes could thrive.

Another possibility in breeding resistance to nematode infestation is the development of resistance from related wild species into the cultivated ones. The related wild species *P. colubrinum* and *P. aduncum* are highly resistant to *M. incognita* (Paulus et al. 1993; Ramana and Mohandas 1986). Additionally, *P. hymenophyllum* and *P. attenuatum* showed remarkable resistance to *R. similis* (Venkitesan and Setty 1978). Though *P. colubrinum* has been found to be immune to *R. similis* (Ramana et al. 1994), crosses between *P. nigrum* and *P. colubrinum* have been unsuccessful until now. The solution for cross transfer of resistance from resistant to cultivated species might be found through the biotechnological route, but, as of now, it still is very challenging. The gene transfer mechanism in pepper is still very poorly understood.

There is also the possibility of employing biological organisms, as in the case of controlling diseases, to control nematode infestation. Sewell (1965) defines biological control as "the induced or natural, direct or indirect limitations of a harmful organism or its effects by another organism or group of organisms." Because a variety of microorganisms inhabit the soil, some of which are either predatory or antagonistic to plant parasitic nematodes, such a course of action could be taken. However, the question of how to biologically control nematodes is still very open. Nevertheless, this approach could be a part of overall integrated pest management in pepper. As of now, the research efforts in the area of biocontrol are, indeed, very sparse. A few examples, given next, illustrate the efficiency of biocontrol.

In the area of biocontrol, fungal agents are the most important. Among them, an opportunistic fungus, namely, *Paecilomyces lilacinus*, has great potential in the control of root knot and cyst nematodes. The fungus, on contact with the egg masses of the nematodes, colonizes and grows rapidly. The chitinolytic enzymes produced by the fungus help penetration of the eggs and cause the suppression of nematodes. The efficacy of the fungus in controlling root knot nematodes has been reported by Jatala (1986). Inoculation of the fungus into rooted cuttings or seedlings of pepper significantly reduced the growth of the root knot nematode, which in turn reduced damage to the roots (Freire and Bridge 1985b; Ramana 1994; Sosamma and Koshy 1995). However, it was less efficient against the control of the borrowing nematode (Geetha 1991; Ramana 1994).

The next fungus that is effective against nematodes is *Trichoderma* spp. It has been found effective in controlling root knot nematode (Eapen and Ramana 1996), in addition to suppressing the *P. capsici*, as detailed earlier in this review. *Verticillium chlamydosporium* is another fungus for control of root knot and cyst nematodes (Kerry 1990). The fungus colonizes the rhizosphere, infects adult females and egg masses, and reduces nematode multiplication by inhibiting the hatching of eggs. Parasitization of root knot nematode eggs by the fungus in Brazil has been reported (Freire and Bridge 1985b). The occurrence of the fungus in association with *Trophotylenchulus piperis* in pepper plantations was first reported in India by Sreeja et al. (1996). In an in vitro test, the fungus was found to colonize the egg masses of *M. incognita* and reduce hatching by almost 50% within 5 days of inoculation (Sreeja et al. 1996). In addition to these, the VAM also has a suppressive effect on nematode population through the indirect effect of enhancing plant vigor, which has been referred to earlier in this review. Occurrence of VAM fungi on pepper roots was first reported in India by Manjunath and Bagyaraj (1982). Bacteria come after fungus as nematicidal agents. Within the bacteria is *Pasteuria penetrans*, which is an obligate parasite of some nematodes. It infects the nematodes by direct penetration through the cuticle by germinating spores sticking to the body surface of the nematodes. It has been found to suppress both *M. incognita* and *R. similis*. In addition, *Bacillus pumilus*, *B. macerans*, and *B. circulans* also suppress *M. incognita* (Sheela et al. 1993). Eapen et al. (1997) reported the inhibitory effect of fluorescent pseudomonas, such as *Pseudomonas fluorescens*, on *M. incognita*. It is important to realize that research in the area of biocontrol of nematodes is in its infant stage; a number of studies are done under *in vitro* conditions, and careful evaluation has to be performed before large-scale field recommendations can be made that are applicable to pepper plantations. The experience, so far, has been promising, though in a limited sense, and much more work needs to be done to probe the area further.

Perhaps the most effective means of controlling nematode infestation is through the use of chemicals. Nematicides—chemicals used to kill nematodes—consist of two groups: systemic nematicides and soil fumigants. Of the two, the latter is more effective because nematodes generally inhabit the soil, and once the nematicide is applied to the soil, it is absorbed by the roots of the host plant and translocated to shoots where they act as systemic inhibitors entering into biochemical reactions within the host plant cell. Despite the effectiveness of nematicides, their widespread use is not in vogue because of high costs and the environmentally adverse fallout from their use. Various nematicides, such as phenamiphos (Nambiar and Sarma 1980), aldicarb sulfone (Venkitesan and Setty 1978), phorate, and DBCP (Venkitesan and Charles 1980), were found to be effective. Among the fumigants, DD, methyl bromide, and ethylene dibromide have been found to be effective. With the advent of organic farming and greater awareness among consumers of products of chemical- free agriculture, use of nematicides is being increasingly restricted. Hopefully, with more research to follow, better and safer nematicides will enter the market.

The next group of pepper pests (following plant pathogenic diseases and plant parasitic nematodes) is infestation by insects. Though a number of insect pests attack the pepper plant, the most devastating is the pollu beetle, *L. nigripennis* Mots—though, to a lesser extent—others, such as top shoot borer (*Cydia hemidoxa* Meyr.), leaf gall thrips (*Liothrips karnyi* Bagn.), and scale insects (*Lepidosaphes piperis* Green and *Aspidiotus destructor* Sign.), could also be considered as pests of major importance. However, this review will primarily focus on the pollu beetle. The insect pest derives the name *pollu* because of the hollow berry that results from its infestation (*pollu* means "hollow" in the regional Malayalam language of Kerala). The status and control of major insect pests in India have been reviewed (Devasahayam et al. 1988; Premkumar et al. 1994).

The extent of attack by the *pollu* beetle varies, and reports of damage vary from 6% to 21% (Thomas and Menon 1939), while Rehiman and Nambiar (1967) report damage as much as 30–40% in the state of Kerala. A survey of one of the major pepper-growing districts of Kerala, the Kannur District (where incidentally the first research station in pepper in the world was established in the town of Panniyur), showed that the single greatest cause for loss in pepper yield was due to infestation by the *pollu* beetle, amounting to as much as 13% (Prabhakaran 1994).

The beetle measures about 2.5 × 1.5 mm and feeds on tender shoots, leaves, and spikes. When infestation is severe, leaves and spikes rot and drop down. The grubs do the most damage. They bore into tender berries and empty their contents; the infested berries in turn become chlorotic first and then turn black and crumble when pressed. Among the pepper cultivars, wide variation has been observed inasmuch as susceptibility or resistance is concerned. Among the four cultivars in which field observations were recorded, *Kalluvally* and *Karimunda* were least susceptible to the pest in the northern and southern regions of the state, respectively (Premkumar and Nair 1988).

Concerning control of the insect pest, both biocontrol and chemical control assume importance in addition to plantation management. Only few natural organisms exist that have a control trait on the pollu beetle. Among these, an unidentified entomophagous nematode (Mermithidae family), a predatory spider (Araneae family), and *Oecophylla smaragdina* Fabr. (Formicidae family) have been identified. The extent of parasitization or predation by these on the pollu beetle is rather limited, and their efficacy in controlling the insect pest in field conditions is open to question (Devasahayam and Koya 1994). Within the ambit of plantation management, application of insecticides is the only viable proposition to effectively control the pollu beetle. A number of insecticides have been evaluated in the field for their efficacy in controlling the pest. On balance, endosulfan spray (0.05%) has been found to be most effective in Kerala (Nandakumar et al. 1987; Premkumar and Nair 1987; Premkumar et al. 1986). It must, however, be noted that of late, there has emerged widespread opposition from environmental activists in the state of Kerala to the use of endosulfan because of suspected human health hazards, such as Mongolian births, central nervous system (CNS) disorders, and even blood cancer. These maladies have been noted in the northern part of Kerala, and the insecticide has been currently banned from use.

Application of insecticides, fungicides, or nematicides leaves pesticide residues in the berries, and with increasing health awareness within the country and outside of it, consumers are opting for organically grown pepper. Though biocontrol is a promising avenue, results obtained so far are of a very preliminary nature, and no hard and fast recommendations have emerged. A number of by-products have been used, both in vitro and in situ in the field. Among them, mention must be made of neem oil, neem seed extract (*Azadirachta indica* A. Juss.), nux-vomica (*Strychnos nux-vomica* L.), and custard apple (*Annona squamosa* L.), and the success rate has not always been very consistent (Devasahayam and Leela 1997), though they show potential for further research (Devasahayam and Anandaraj 1997).

As far as Malaysia and Indonesia are concerned, the pepper weevil (*Lophobaris piperis*) is the most serious insect pest. The grubs bore into the nodal region of the climbing and flowering shoots, which result in infested plants. As the infestation advances, the entire aerial portion of the vine wilts and collapses. The infestation of the pest in the lower part of the vine results in the most damage, amounting to a yield loss of approximately 5–50% in the upper part (Deciyanto and Suprapto 1992). A number of natural predators have been identified, of which *Spathius piperis* has been found to be the most effective in controlling the pest both in the laboratory and in the field. As much as 40% control has been reported (Deciyanto and Suprapto 1992). A number of insecticides have been recommended for the control of the pepper weevil. These include the spraying of endosulfan (0.2%), parathion (0.2%), or permethrin (0.2%). Removal of affected branches and stems would also help in reducing the level of pest infestation in the field (Deciyanto and Suprapto 1992; Kueh 1979).

References

Ahmed K (1993) The effect of Nitrogen on the growth of *Piper nigrum* L. planted with *Desmodium trifolium* cover. In: Ibrahim MY, Bong CFJ, Ipor IP (eds) The pepper industry: problems and prospects. Universiti Pertanian Malaysia, Sarawak, pp 97–105

Anandaraj M (1997) Ecology of *Phytophthora capsici* (Leonian 1922, emend A. Alizadeh and P.H. Tsao). In: Causal organism of foot rot disease of black pepper (*Piper nigrum* L.). Ph. D. thesis, University of Calicut, Kerala State, India, p 154

Anandaraj M, Peter KV (1996) Biological control in spices. Indian Institute of Spices Research, Calicut, p 52

Anandaraj M, Sarma YR (1994a) Effect of Vesicular Arbuscular Mycorrhiza on rooting of black pepper (*Piper nigrum* L.). J Spices Aromatic Crops 3:39–42

Anandaraj M, Sarma YR (1994b) Biological control of black pepper diseases. Indian Cocoa Arecanut Spices J 18:22–23

Anandaraj M, Sarma YR (1995) Diseases of black pepper (*Piper nigrum* L.) and their management. J Spices Aromatic Crops 4:17–23

Anandaraj M, Sarma YR (1997) Mature coconut water for mass culture of biocontrol agents. J Plant Crops 25:112–114

Anandaraj M, Sarma YR, Ramachandran N (1994) *Phytophthora* root rot of black pepper in relation to age of the host and its culmination in foot rot. Indian Phytopathol 47:203–206

Biessar S (1969) Plant parasitic nematodes of crops in Guyana. PANS 15:74–75

Bopaiah BM, Khader KBA (1989) Effect of bio-fertilizers on growth of black pepper (*Piper nigrum* L.). Indian J Agric Sci 59:682–683

Bridge J (1978) Plant nematodes associated with cloves and black pepper in Sumatra and Bangka, Indonesia. In: O.D.M. Technical Report, U.K. Ministry of Overseas Development, UK, p 19

Christie JR (1959) Plant nematodes—their bionomics and control. University of Florida, Gainesville, p 256

Coffey MD (1991) Strategies for the integrated control of soil-borne *Phytophthora* species. In: Lucas JA, Shattock RC, Shaw DS, Cooke LR (eds) Phytophthora. Cambridge University Press, Cambridge, pp 411–432

Cook RJ, Baker R (1983) The nature and practice of biological control of plant pathogens. American Phytopathological Society, St. Paul, p 53

Das TPM, Cheeran A (1986) Infectivity of *Phytophthora* species on cash crops in Kerala. Agric Res J Kerala 24:7–13

Datta PK (1984) Studies on two *Phytophthora* diseases (*Koleroga* of arecanut and black pepper wilt) in Shimoga District, Karnataka State, India. Ph.D. thesis. University of Agricultural Sciences, College of Agriculture, Dharwad, Karnataka State, India, p 121

Davis RM, Menge JA (1981) *Phytophthora parasitica* inoculation and intensity of vesicular–arbuscular mycorrhizae in citrus. New Phytol 87:705–715

Davis RM, Menge JA, Zentmyer GA (1978) Influence of vesicular arbuscular mycorrhiza on *Phytophthora* root rot of three crop plants. Phytopathology 68:1614–1617

Deciyanto S, Suprapto (1992) Research progress on important insect pests of black pepper in Indonesia. In: Wahid P, Sitepu D, Deciyanto D, Suparman U (eds) Proceedings of the international workshop on black pepper diseases, 3–5 December, 1991, Bandar Lampung, Indonesia. Research Institute for Spice and Medicinal Crops, Bogor, pp 237–244

Dehne HW (1982) Interaction between vesicular–arbuscular mycorrhizal fungi and plant pathogens. Phytopathology 72:1115–1119

Devasahayam S, Anandaraj M (1997) Black pepper (*Piper nigrum* L.) research. In: Narwal SS, Tauro P, Bisla SS (eds) Neem in sustainable agriculture. Scientific Publishers, Jodhpur, pp 117–122

Devasahayam S, Koya KMA (1994) Natural enemies of major insect pests of black pepper (*Piper nigrum* L.) in India. J Spices Aromatic Crops 3:50–55

Devasahayam S, Leela NK (1997) Evaluation of plant products for antifeedant activity against pollu beetle (*Longitarsus nigripennis* Motschulsky), a major pest of black pepper Abst. First national symposium on pest management in horticultural crops. Environmental implications and thrusts, Bangalore, p 85

Devasahayam S, Premkumar T, Koya KMA (1988) Insect pests of black pepper *Piper nigrum* L. in India—a review. J Plant Crops 16:1–11

Drew MC, Lynch JM (1980) Soil anaerobiosis, microorganisms and root function. Annu Rev Phytopathol 18:37–67

Eapen SJ, Ramana KV (1996) Biological control of plant parasitic nematodes of spices. In: Anandaraj M, Peter KV (eds) Biological control in spices. Indian Institute of Spices Research, Calicut, pp 20–32

Eapen SJ, Ramana KV, Sarma YR (1997) Evaluation of *Psuedomonas fluorescens* iso- lates for control of *Meloidogyne incognita* in black pepper (*Piper nigrum* L.). In: Edison S, Ramana KV, Sasikumar B, Nirmal Babu K, Eapen SJ (eds) Biotechnology of spices, medicinal and aromatic plants. Indian Society for Spices, Calicut

Ewald S (1991) Vesicular–arbuscular mycorrhiza management in tropical agrosystems. Technical Cooperation, Federal Republic of Germany, p 371

Fassuliotis G, Bhatt DP (1982) Potential of tissue culture for breeding root knot nematode resistance in vegetables. J Nematol 14:10–14

Ferraz ECA, Orchard JE, Lopez AS (1984) Reactions of black pepper to *Meloidogyne incognita* in relation to total phenol. Revista Theobroma 14:217–227

Ferraz ECA, Lordello LGE, de Santana CJL (1988) Nutrient absorption of black pepper vine (*Piper nigrum* L.) infested with *Meloidogyne incognita* Kofoid and White., 1919, Chitwood (1949). In: Boletin Tecnico Centro de Pesquisas do Cacau, Brazil, No. 160, p 34

Ferraz ECA, Lordello LGE, Gonzaga E (1989) Influence of *Meloidogyne incognita* (Kofoid and White 1919) Chitwood 1949 on chlorophyll content of black pepper (*Piper nigrum* L.). Agrotropica 1:57–62

Freire FCO, Bridge J (1985a) Biochemical changes induced in roots and xylem sap of black pepper by *Meloidogyne incognita*. Fitopatologia Brasileira 10:483–497

Freire FCO, Bridge J (1985b) Parasitism of eggs, females of juveniles of *Meloidogyne incognita* by *Paecilomyces lilacinus* and *Verticillium chlamydosporium*. Fitopatologia Brasileira 10:577–596

Geetha SM (1991) Studies on the biology, pathogenicity and biocontrol of different populations of *Radopholus similis.* Ph.D. thesis, Kerala University, Trivandrum, India, p 196

Govindan M, Chandy KC (1985) Utilization of the diazotroph, *Azospirillum* for inducing rooting in pepper cuttings (*Piper nigrum*). Curr Sci 54:1186–1188

Graham JH (1982) Effect of citrus root exudates on germination of chlamydospores of the vesicular–arbuscular mycorrhizal fungus *Glomus epigaeum*. Mycologia 74:831–835

Graham JH (1988) Interactions of mycorrhizal fungi with soil borne plant pathogens and other organisms: an introduction. Phytopathology 78:365–366
Graham JH, Egel DS (1988) *Phytophthora* root rot development on mycorrhizal and phos- phorus-fertilized non-mycorrhizal sweet orange seedlings. Plant Dis 72:614–615
Holdeman QL (1986) The Burrowing Nematode, *Radopholus similis* sensu lato. Division of Plant Industry, Department of Food and Agriculture and the Agriculture Commissioners of California, California, USA, p 52
Holliday P, Mowat WP (1963) Foot rot of *Piper nigrum* L. (*Phytophthora palmivora*). In: Phytopathological Paper no. 5. Commonwealth Mycological Institute, Kew/Surrey, p 62
Hubert FP (1957) Diseases of some export crops in Indonesia. Plant Dis Report 41:55–63
Ichinohe M (1975) Infestation of black pepper vines by the root knot nematode, *Meloidogyne incognita* at Tome-Acu, Para, Brazil. Jpn J Nematol 5:36–40
Ichinohe M (1980) Studies on the root knot nematode of black pepper plantation in Amazon. In: Annual report of the Society of Plant Protection of North Japan, No. 31, pp 1–8
Ichinohe M (1985) Integrated control of root knot nematode, *Meloidogyne incognita* on black pepper plantations in the Amazonian region. Agric Ecosyst Environ 12:271–283
Jacob JA, Kuriyan KJ (1979) Screening of pepper varieties for resistance against root knot nematode (*Meloidogyne incognita*). Agric Res J Kerala 17:90
Jatala P (1986) Biological control of plant parasitic nematodes. Annu Rev Phytopathol 24:453–489
Kerry BR (1990) An assessment of progress towards microbial control of plant parasitic nematodes. J Nematol 22(Suppl):621–631
Koshy PK, Sunderaraju P (1979) Response of seven black pepper cultivars to *Meloidogyne incognita*. Nematol Medit 7:123–125
Kueh TK (1979) Pests, diseases and disorders of black pepper in Sarawak Semongok Agricultural Research Centre. Department of Agriculture, Sarawak, p 68
Kueh TK (1986) Pests and diseases of black pepper – a review. In: Bong CFJ, Saad MS (eds) Pepper in Malaysia. Universiti Pertanian Malaysia, Sarawak, pp 115–133
Kueh TK (1990) Major diseases of black pepper and their management. Planter 66:59–69
Kueh TK, Khew KL (1982) Survival of *Phytophthora palmvivora* in soil and after passing through alimentary canals of snails. Plant Dis 66:897–899
Kueh TK, Sim SL (1992a) Etiology and control of *Phytophthora* foot rot of black pep- per in Sarawak, Malaysia. In: Wahid P, Sitepu D, Deciyanto S, Suparman U (eds) Proceedings of the International workshop on black pepper diseases. Bander Lampung, Indonesia. Institute for Spice and Medicinal Crops, Bogor, pp 155–162
Kueh TK, Sim SL (1992b) Important pepper diseases in Sarawak, Malaysia. In: Wahid, P., Sitepu, D., Deciyanto, S., Suparman, U. (eds.), Proceedings of the international workshop on black pepper diseases Bander Lampung, Bogor, Indonesia, pp 101–117
Kueh TK, Fatimah O, Lim JL (1993) Management of black berry disease of black pepper. In: Ibrahim MY, Bong CFJ, Ipor IB (eds) The pepper industry: problems and prospects. Universiti Pertanian Malaysia, Sarawak, pp 162–168
Linderman RG (1988) Mycorrhizal interactions with the rhizosphere microflora: the micorrhizosphere effect. Phytopathology 78:366–370
Lokesh MS, Gangadarappa PM (1995) Management of *Phytophthora* foot rot and nematode diseases in black pepper (*Piper nigrum* L.). J Spices Aromatic Crops 4:61–63
Malebennur NS, Ganadarappa PM, Hegde HG (1991) Chemical control of foot rot of black pepper caused by *Phytophthora capsici* (*Phytophthora palmivora* MF 4). Indian Cocoa Arecanut Spices J 14:148–149
Mamootty KP (1978) Quick wilt disease of pepper (*Piper nigrum* Linn.)-1: symtomatological studies on the quick wilt disease of pepper. M.Sc. (Ag.) thesis, Kerala Agricultural University, Thrissur, Kerala State, India, p 87
Manjunath A, Bagyaraj DJ (1982) Vesicular arbuscular mycorrhizas in three plantation crops and cultivars of field bean. Curr Sci 51:707–709

Manohara D, Kasim R, Sitepu D (1992) Current research status of foot rot disease in Indonesia. In: Wahid P, Sitepu D, Deciyanto S, Suparman U (eds) Proceedings of the international workshop on black pepper diseases Bander, Lampung, Indonesia. Institute for Spice and Medicinal Crops, Bogor, pp 144–154

Menon KK (1949) The survey of pollu and root diseases of pepper. Indian J Agric Sci 19:89–136

Mohandas C, Ramana KV (1987) Slow wilt disease of black pepper and its control. Indian Cocoa Arecanut Spices J 11:10–11

Mohandas C, Ramana KV (1991) Pathogenicity of Meloidogyne incognita and Radopholus similis on black pepper (*Piper nigrum* L.). J Plant Crops 19:41–43

Mustika I (1990) Studies on the interactions of *Meloidogyne incognita, Radopholus similis* and *Fusarium solani* on black pepper (*Piper nigrum* L.). Ph.D. thesis, Wageningen Agricultural University, Wageningen, The Netherlands, p 127

Mustika I (1991) Response of four black pepper cultivars to infection by *Radopholus similis*, *Meloidogyne incognita* and *Fusarium solani*. Ind Crop Res J 4(1):17–22

Mustika I, Zainuddin N (1978) Efficacy tests of some nematicides for the control of nematodes on black pepper. Pemberitaan LPTI 30:1–10

Nair PKU, Sasikumaran S (1991) Effect of some fungicides on quick wilt (foot rot) disease of black pepper. Indian Cocoa Arecanut Spices J 14:95–96

Nair MRGK, Das NM, Menon MR (1966) On the occurrence of the burrowing nematode, *Radopholus similis* (Cobb 1893) Thorne 1949, on banana in Kerala. Indian J Entomol 28:553–554

Nair PKU, Mammootty KP, Sasikumaran S, Pillay VS (1993) *Phytophthora* foot rot of black pepper (*Piper nigrum* L.). A management study with organic amendments. Indian Cocoa Arecanut Spices J 17:1–2

Nambiar KKN (1978) Diseases of pepper in India. In: Nair MK, Haridasan M (eds) Proceedings of the national seminar on pepper. Central Plantation Crops Research Institute, Calicut, pp 11–14

Nambiar KKN, Sarma YR (1977) Wilt diseases of black pepper. J Plant Crops 5:92–103

Nambiar KKN, Sarma YR (1980) Factors associated with slow wilt of pepper. Proceedings of the PLACROSYM-II 1979. Indian Society for Plantation Crops, Kasaragod, pp 348–358

Nandakumar C, Ragunath P, Visalakshi A, Mohandas N (1987) Management of pollu beetle on pepper by adjusting the spray schedules. In: Abstracts of national symposium integrated pest control: progress and perspectives, Trivandrum, Kerala State, India, p 50

Paulus AD, Eng L, Teo CH, Sim SL (1993) Screening black pepper genotypes and *Piper* spp. for resistance to root knot nematode. In: Ibrahim MY, Bong CFJ, Ipor IP (eds) The black pepper industry: problems and prospects. Universiti Pertanian Malaysia, Sarawak, pp 132–139

Prabhakaran PV (1994) Assessment of losses due to pests, diseases and drought in black pepper. J Indian Soc Agric Stat 47:59–60

Premkumar T, Nair MRGK (1987) Reduction in damage by *Longitarsus nigripennis* to berries of black pepper by insecticides. Madras Agric J 74:326–327

Premkumar T, Nair MRGK (1988) Effect of some planting conditions on infestation of black pepper by *Longitarsus nigripennis* Mots. Indian Cocoa Arecanut Spices J 10:83–84

Premkumar T, Banerjee SK, Devasahayam S, Koya KMA (1986) Effect of different insecticides on the control of pollu beetle *Longitarsus nigripennis* Mots.: a major pest of black pepper *Piper nigrum* L. Entomon 11:219–221

Premkumar T, Devasahayam S, Koya KMA (1994) Pests on spice crops. In: Chadha KL, Rethinam P (eds) Advances in horticulture: plantation and spice crops, Part 2, vol 10. Malhotra Publishing House, New Delhi

Ramachandran N (1990) Bioefficacy of systemic fungicides against Phytophthora infections in black pepper (*Piper nigrum* L.). Ph.D. thesis, University of Calicut, Calicut, Kerala State, India, p 138

Ramachandran N, Sarma YR (1985) Efficacy of three systemic fungicides in controlling Phytophthora infections in black pepper. Indian Phytopathol 38:160–162

Ramachandran N, Sarma YR, Nambiar KKN (1986) Spatial distribution of *Phytophthora palmivora* MF4 in the root zones of *Piper nigrum*. Indian Phytopathol 39:414–417

Ramachandran N, Sarma YR, Anandaraj M (1988a) Sensitivity of Phytophthora species affecting different plantation crops in Kerala to Metalaxyl. Indian Phytopathol 41:438–442

Ramachandran N, Sarma YR, Anandaraj M, Abraham J (1988b) Effect of climatic factors on *Phytophthora* leaf infection in black pepper grown in arecanut–black pepper mixed cropping system. J Plant Crops 16:110–118

Ramachandran N, Sarma YR, Anandaraj M (1990) Vertical progression and spread of *Phytophthora* leaf infection in black pepper in arecanut–black pepper mixed cropping system. Indian Phytopathol 43:414–419

Ramachandran N, Sarma YR, Anandaraj M (1991) Management of *Phytophthora* infections in black pepper. In: Sarma YR, Premkumar T (eds) Diseases of black pepper. National Research Centre for Spices, Calicut, pp 158–174

Ramana KV (1992) Final report of the project: role of nematodes in the incidence of slow decline (slow wilt disease) of black pepper and screening pepper germplasm against nematodes. National Research Centre for Spices, Calicut, p 149

Ramana KV (1994) Efficacy of *Paecilomyces lilacinus* (Thom.) Samson in suppressing nematode infestations in black pepper (*Piper nigrum* L.). J Spices Aromatic Crops 3:130–134

Ramana KV, Eapen SJ (1998) Plant parasitic nematodes associated with spices and condiments. In: Trivedi PC (ed) Nematode diseases in plants. C. B. S. Publishers and Distributors, New Delhi, pp 217–251

Ramana KV, Mohandas C (1986) Reaction of black pepper germplasm to root knot nematode *Meloidogyne incognita*. Indian J Nematol 16:138–139

Ramana KV, Mohandas C, Ravindran PN (1987) Reaction of black pepper germplasm to the burrowing nematode (*Radopholus similis*). J Plant Crops 15:65–66

Ramana KV, Mohandas C, Eapen SJ (1994) Plant parasitic nematodes and slow decline disease of black pepper. Technical Bulletin National Research Centre for Spices, Calicut, p 14

Ramesh CR (1982) Root infection and population density of VA mycorrhizal fungi in a coconut based multi-storied cropping. In: Bavappa KVA, Iyer RD, Das PK, Sethuraj MR, Ranganathan V, Nambiar KKN, Chacko MJ (eds) Proceedings of the Plantation Crops Symposium (PLACRO-SYM V). Central Plantation Crops Research Institute, Kasaragod, pp 548–554

Rehiman PVA, Nambiar PKV (1967) Insecticidal control of pollu beetle (*Longitarsus nigripennis*) on pepper. Madras Agric J 54:39–40

Sadanandan AK, Abraham J, Anandaraj M (1992) Impact of high production technology on productivity and disease incidence of black pepper. In: Nambiar KKN, Iyer RD, Rao EVVB (eds) Proceedings of the PLACROSYM-IX, 20. Indian Society of Plantation Crops, Kasaragod

Samraj J, Jose PC (1966) A Phytophthora wilt of pepper (*Piper nigrum*). Sci Cult 32:90–92

Sarkar BB, Choudhary DM, Nath SC, Bhattacharjee NR, Barna DTC (1985) Wilt disease of pepper and its control in Tripura. Indian Cocoa Arecanut Spices J 8:63–66

Sarma YR, Anandaraj M, Ramana KV (1992) Present status of black pepper diseases in India and their management. In: Wahid P, Sitepu D, Deciyanto S, Suparman U (eds) Proceedings of the international workshop on black pepper diseases Bander Lampung, Indonesia. Institute for Spice and Medicinal Crops, Bogor, pp 67–68

Sarma YR, Anandaraj M, Venugopal MN (1996) Biological control of diseases in spices. In: Anandaraj M, Peter KV (eds) Biological control in spices. Indian Institute of Spices Research, Calicut, pp 1–19

Sewell GWF (1965) The effect of altered physical conditions of soil on biological control. In: Baker KF, Snyder WC (eds) Ecology and soil borne plant pathogens. University of California Press, Berkeley, pp 479–494

Sharma RD, Loof PAA (1974) Nematodes of cocoa region of Bahia, Brazil IV. Nematodes in the rhizosphere of pepper (*Piper nigrum* L.) and clove (*Eugenia caryophylla* Thumb). Revista Theobroma 4:26–32

Sheela MS, Venkitesan TS, Mohandas N (1993) Status of *Bacillus* spp. as biocontrol agents of root knot nematode (*Meloidogyne incognita*) infesting black pepper (*Piper nigrum* L.). J Plant Crops 21(Suppl):218–222

Sher SA, Chunram C, Pholcharoen S (1969) Pepper yellows disease and nematodes in Thailand. FAO Plant Prot Bull 17:33

Sitepu D, Kasim R (1991) Black pepper diseases in Indonesia and their control strategy. In: Sarma YR, Premkumar T (eds) Diseases of black pepper. National Research Centre for Spices, Calicut, pp 13–28

Siti Hajijah AS (1993) Observations of root knot infestation on pepper (*Piper nigrum* L.) in Sarawak. In: Ibrahim MY, Bong CFJ, Ipor IB (eds) The pepper industry: problems and prospects. Universiti Pertanian Malaysia, Sarawak, pp 140–147

Sosamma VK, Koshy PK (1995) Effect of *Pasteuria penetrans* and *Paecilomyces lilacinus* on population build up of root knot nematode, *Meloidogyne incognita* on black pepper. Indian J Nematol 25:16–17. (Abst.)

Sreeja TP, Eapen SJ, Ramana KV (1996) Occurrence of *Verticillium chlamydosporium* Goddard in a black pepper (*Piper nigrum* L.) garden in Kerala, India. J Spices Aromat Crops 5:143–147

Stirling GR (1991) Biological control of plant parasitic nematodes. C. A. B. International, Wallingford, p 282

Sundararaju P, Koshy PK, Sosamma VK (1979) Plant parasitic nematodes associated with spices. J Plant Crops 7:15–26

Sundararaju P, Koshy PK, Sosamma VK (1980) Survey of nematodes associated with spices in Kerala and Karnataka. In: Proceedings of the PLACROSYM-II, 1979. Indian Society for Plantation Crops, Kasaragod, pp 39–44

Thomas KM, Menon KK (1939) The present position of pollu disease of pepper in Malabar. Madras Agric J 17:347–356

Van der Vecht J (1950) Plant parasitic nematodes. In: Karshoven LGE, van der Vecht J (eds) Diseases of cultivated plants in Indonesian Colonies. W. van Hoeve, I. S' gravenhage, pp 16–45

Varughese J, Anuar MA (1992) Etiology and control of slow wilt disease in Johore, Malaysia. In: Wahid P, Sitepu D, Deciyanto S, Suparman U (eds) Proceedings of the international workshop on black pepper diseases. Research Institute for Spice and Medicinal Crops, Bogor, pp 188–197

Venkitesan TS, Charles JS (1980) A note on the chemical control of nematodes infesting pepper vines in Kerala. In: Proceedings of the PLACROSYM-II 1979. Indian Society for Plantation Crops, Kasaragod, pp 27–30

Venkitesan TS, Setty KGH (1978) Reaction of 27 black pepper cultivars and wild forms to the burrowing nematode *Radopholus similis* (Cobb) Thorne. J Plant Crops 6:81–84

Wahid P (1976) Studies on yellows disease in black pepper on the island of Bangka. Pembr LPTI 21:64–79

Winoto SR (1972) Effect of *Meloidogyne* species on the growth of *Piper nigrum* L. Malaysian Agric Res 1:86–90. Seoul Teahakkyo Saengyak Yonguso Opidukdip

Zadocks JC, Van den Bosch F (1994) On the spread of plant disease: a theory of foci. Annu Rev Phytopathol 32:503–521

Chapter 7
The Processing of Black Pepper on Farm

Abstract The chapter primarily discusses important aspects of black pepper processing. It discusses harvest schedule in relation to end products in pepper and changes in chemical composition of two important pepper cultivars, namely, Panniyur-1 and Karimunda. The chapter also discusses the details of solar drying of pepper, garbling, cleaning and fractionation, packaging, and storing.

Keywords Sun drying · Solar drying · Garbling · Harvest schedule · End products · Chemical composition

The processing of pepper starts from the harvesting operation, which is a long, drawn-out process. With the onset of the southwest monsoon in the months of May and June, the pepper plants start to flower. The harvest commences in December and spills over into January of the following year, when there is ample sunshine and the rains completely cease. Varietal differences and the prevalent climatic pattern decide the harvesting process. In the plains of the state of Kerala, the early maturing varieties reach the harvest stage by November and the late maturing ones by December–January. Harvesting of pepper is important in view of the end products, which have a very great industrial significance. Govindarajan (1979) has made a detailed and systematic analysis of the harvesting dates against the background of the end products as detailed in Tables 7.1 and 7.2. Data in Table 7.2 give a clear idea how the chemical composition varies in relation to maturity dates.

Harvesting commences based mainly on the visual appearance of the berries. When a couple of berries in a spike turn orange or red in color, harvesting commences. Only the mature spikes are harvested. When dry, the immature spikes tend to produce shriveled berries, and this in turn will reduce the quality of the produce. However, in Sri Lanka, premature harvesting is done, and the produce is generally known as "light pepper" (Purseglove et al. 1981), which garners a premium in the market because it contains a higher amount of oleoresin and volatile oils. Harvesting is done manually by climbing up the vine using a bamboo pole; the produce is collected in a jute bag tied to the back of the climber. Only in flat

K. P. Nair, *The Geography of Black Pepper (*Piper nigrum*)*,
https://doi.org/10.1007/978-3-030-52865-2_7

Table 7.1 Harvest schedule in relation to the end products of pepper

End products	Time to harvest
Black pepper	Fully mature and nearly ripe
White pepper	Fully ripe
Canned pepper	4- to 5-month-old berries
Dehydrated green pepper	10–15 days before full maturity
Oleoresin	15–20 days before full maturity
Pepper oil	15–20 days before full maturity
Pepper powder	Fully mature berries when starch content peaks

Table 7.2 Changes in chemical composition of two important cultivars in relation to maturity dates

	Maturity date									
	Months after fruit setting									
Cultivar	Panniyur-1					Karimunda				
End products	3.0	4.5	5.5	6.5	7.5	3.0	4.5	5.5	6.5	7.0
Volatile oil (%)	6.4	7.6	6.3	2.8	2.0	6.8	10.4	8.2	4.4	3.6
NVEE[a] (%)	8.7	8.8	8.7	8.1	7.8	10.3	9.7	8.6	7.5	7.4
Piperine content (%)	1.9	2.6	2.7	3.1	3.5	1.9	2.4	2.4	3.1	3.1
Starch content (%)	2.5	3.7	5.1	10.2	16.8	2.6	4.9	6.2	15.3	15.3

[a]*NVEE* Nonvolatile ether extract

plantations can mechanical ladders attached to trolleys be used for harvesting. In India, pepper plantations are usually on slopy terrain, and such harvesting can only be done by manual labor.

Once harvesting is completed and the produce brought to the farmyard, post-harvest operations have to begin. Proper handling of the post-harvest produce is very crucial to maintain the quality of the produce.

The first step in post-harvest handling is "decorning" or removal of the berries from the stalk. It is synonymous to the threshing of wheat or rice grains. In most situations, especially homestead farming and large-scale plantations, decorning is still done manually by spreading the produce on a clean floor in the farmyard, where laborers, mainly women, use their feet to trample the spikes to separate the berries from the stalk. It is not very hygienic, yet in most Indian situations, there seems to be no alternative. Mechanical threshers are a rarity, though they are in use in Indonesia and Malaysia. When mechanical threshers are used, the green spikes are slowly fed into the thresher, in which a rotating drum with aluminum blades removes the berries from the stalks. The speed of the rotation is adjusted to be neither too slow nor too fast. The spikes are fed into a moving drum with aluminum plates with suitable gaps; due to the mild stretching action, the berries are separated from the stalk. A thresher powered by a 3-hp motor can handle 1.5 tons/h. Before decorning, it should be ensured that the spikes are properly washed clean of all adhering dirt. Once the berries are threshed, they should be graded using a mesh; normally, three grades are in vogue, namely, 4.25 mm (large), 3.25–4.25 mm (medium), and 3.25 mm (small). The first released hybrid Panniyur-1 belongs to the large size.

After grading, the berries are blanched by dipping the berries carried in bamboo baskets in boiling water contained in troughs. This is done to clean the produce of all adhering dust particles and any other extraneous particles, such as bird excreta. The produce is then dried to a uniform moisture content (8–10%) in open sunlight for 3 consecutive days. In India, blanching is done in boiling water for 1 min; in Papua New Guinea, Indonesia, and Micronesia, it is done at 82 °C for 2 min (Pruthi 1992). The entire operation of washing, blanching, and grading can be mechanized by spraying boiling water onto pepper fruits as they move in a conveyor belt fitted with mechanical brushes; the blanching time is adjusted through the speed of the conveyor belt. At the end of the conveyor belt, the blanched and washed berries pass through sieves of various dimensions to separate "pinhead" (good and well-sized berries) and light pepper. The graded pepper is later flushed with dry air to remove surface moisture and then transported to the drying yard or to artificial driers.

The phenolase enzyme, which imparts the black color to the berries, is activated by blanching. It also ruptures the cells and thereby accelerates the escape of moisture from the inner core, simultaneously enhancing the black color with the resinoids inside the berry. Therefore, blanched pepper will shine more and dries at a faster rate. The black color the berries acquire on drying is due to the oxidation of colorless phenolic compounds present in the skin. Polyphenolase (0-diphenol oxidase) present in the fruit wall converts colorless phenolic substrates (3,4-dihydroxy phenyl ethanol glycoside) present in the cells to black polymeric compounds (Variyar et al. 1988).

7.1 Sun Drying of Pepper

Sun drying reduces the initial moisture content in the berries from around 65% to approximately 10%. Most of the pepper-growing countries practice sun drying. To check the development of mold, the berry heap, which is spread on a clean floor for drying, is periodically turned over. On average, sun drying leads to a recovery of 29–38% of the moisture content. Nonuniformity and contamination by microorganisms are the major disadvantages of sun drying. Pepper is dried on different surfaces, such as bamboo mats; cement floors; or black, low-density, polyethylene sheets. It has been observed that the black surface of the sheet, due to absorption and retention of heat, results in faster drying than when other surfaces are used (Krishnamoorthy and John Zachariah 1992). In the Wayanad District in Kerala, a special type of drying is practiced where after 2 days of sun drying on bamboo mats or a cement floor, the dried berries are collected in jute sacks and stacked upright for 2 days in rooms. During this process of storing, a fermentation process begins that imparts a uniform black color to the berries, leading to enhanced flavor. After 2 days of storage the berries are, once again, spread on the cement floor or bamboo mats and dried in the conventional manner. It is for this special type of drying that the pepper of this region is highly regarded for its unique quality of color and flavor.

7.2 Solar Drying of Pepper

Solar dryers have been developed to dry agricultural products in India (Kachru and Gupta 1993) at the Central Institute for Agricultural Engineering in India and in Germany at the Institute of Agricultural Engineering for the Tropics and Subtropics, University of Hohenheim (Esper and Muhlbauer 1996). In the solar dryer, the change occurs mainly in the final drying phase, whereas sun drying requires several days to reach the desired moisture level. Solar drying accelerates the drying process, leading to a considerable reduction in drying time. Additionally, the maintenance of a constant temperature throughout the drying process ensures a clear product free of microbial contamination or any other extraneous matter.

Drying of berries can also be done using mechanical dryers. Different types are in use, such as the "copra dryer," which is used to dry coconut shells and husks; a convection dryer, which is a forced draft dryer; and a cascade-type dryer using indirect heating by kerosene or gas.

After drying, moisture recovery normally does not exceed 70%. Dry recovery varies from 29% to 38%, depending on the variety. Table 7.3 shows the composition of dried pepper.

7.3 Garbling, Cleaning, and Fractionation

Dried pepper is cleaned to get rid of dirt, grit, stone, leaves, and any other extraneous matter before packing. Pneumatic separators equipped with magnetic separators are used to remove metallic contaminations, such as iron filings, nails, and the like. For destoning of spices, a combination of air classification and vibratory conveying using inclined docks is very efficient (Ramanathan and Rao 1974). Some of the well-established processing houses clean and grade pepper with the help of multiple sieve-*cum*-air classifier machines, whereby dust, stalks, pinheads, hollows,

Table 7.3 Chemical composition of dried pepper (in percentage)

Chemical constituent	Composition
Moisture	8.7–14.0
Total nitrogen	1.5–2.6
Volatile ether extract	0.3–4.2
Nonvolatile ether extract	3.9–11.5
Alcohol extract	4.4–12.0
Starch (acid hydrolysis)	28.0–49.0
Crude fiber	8.7–18.0
Crude piperine	2.8–9.0
Piperine	1.7–7.4
Total ash	3.6–5.7
Acid insoluble ash	0.03–0.55

Source: After Pruthi (1993)

immature pepper, red pepper, and extra bold pepper are removed (Pruthi 1992, 1993). Separated pepper is then washed and dried to make it free of adhering fungus and other extraneous matter. Hence, to obtain good-quality pepper of international standard, the following steps are necessary: (1) drying, (2) separation of various fractions, (3) size segregation or grading, (4) physical cleaning (washing and drying), and (5) packing. In India, farmers as well as intermediaries bring the produce to the market, usually in an ungraded form, and garbling and grading are generally performed at the exporters' premises. Most of the exporters clean and grade using garbling machines, which remove dust and chaff and grade the pepper according to densities. However, most of the machines do not remove iron filings and heavy metals. These are generally handpicked. Pepper collected by the large merchants (exporters) is first cleaned manually by picking out the contaminants, which is mostly done by women. After this, the pepper goes for garbling. A spiral separator has been developed by Madasamy and Gothandapani (1995) that ideally removes all extraneous materials. Traditionally, pepper is manually cleaned by winnowing and then packed into jute bags. On some plantations, cleaning is done through a funnel-like arrangement in which dried pepper is fed and cleaned using a blower at the outlet point when the pinheads and light berries are blown away.

The Directorate of Marketing and Inspection, Government of India, has classified pepper into the following grades:

1. Malabar garbled (MG grades 1 and 2)
2. Malabar ungarbled (MUG grades 1 and 2)
3. Tellicherry garbled black pepper special extra bold (TGSEB)
4. Tellicherry garbled extra bold (TGEB)
5. Tellicherry garbled (TG)
6. Garbled light pepper (GL special, GL grades 1 and 2)
7. Ungarbled light pepper (UGL special, UGL grades 1 and 2)
8. Pinheads (PH grade special and PH grade 1)
9. Black pepper (nonspecified) (NS grade X)

Graded pepper fetches a higher price than ungraded, and in the international market, TGSEB fetches a premium. Good garbled pepper should have a bulk density of 500–600 g l^{-1}. Light berries should be less than 10% and pinheads (unfertilized berries) less than 4%. When bulk density is low, it indicates the presence of more lightweight berries, with subsequently less starch content, leading to poor milling quality. Only when the end product has a good aroma and a biting taste will it be considered good. The end product should contain at least 1.5% volatile oil and 3% piperine (Lewis 1984). Table 7.4 contains data pertaining to the physicochemical composition of different grades of black pepper.

There is another grade of pepper known as "half pepper," which is placed between normal pepper and light pepper. The berries are slightly undermature and, therefore, contain more of the active principle piperine. Due to immaturity, the berries may appear slightly wrinkled, and this is ideal for the extraction of oleoresin. The concentration of piperine in oleoresin is more important than the aroma, which is determined by the oil content.

Table 7.4 Physicochemical composition of different grades of black pepper (in percentage)

Grade	Moisture	Volatile oil	Piperine	NVEE[a]	Starch	Crude fiber
MG	13.0	3.7	5.0	12.3	39.7	11.8
MUG	12.0	2.8	5.0	11.4	41.8	12.5
TGSEB	10.0	3.2	4.9	10.3	40.9	9.2
TGEB	13.0	2.2	4.4	9.1	39.7	11.8
Light pepper	13.0	2.9	4.1	13.5	14.6	27.8
Tellicherry ungarbled	12.0	4.0	6.3	13.5	39.3	11.0
Pin heads	13.0	0.6	0.8	7.1	11.5	27.4
High-range ungarbled	12.0	2.6	4.0	11.1	41.8	10.5

[a]*NVEE* Nonvolatile ether extract

Using appropriate sieves, the pepper is graded into different sizes. Very small and underdeveloped berries are classified as pinheads. The bulk of the pepper belongs to the average size, known as Malabar garbled. The larger-sized ones are classified as Tellicherry bold, Tellicherry extra bold, Tellicherry special extra bold, and giant (Kachru et al. 1990; Mathew 1992).

In normal pepper, starch accounts for 34%, while it is 56.5% in white pepper and 63.2% in decorticated white pepper. Of the 12% water-soluble nitrogen present in the berries, nonprotein nitrogen constitutes about 82%, and of this, more than 50% is made up of simple amino acids that act as a nutrient for humans when pepper is consumed as a culinary material.

One of the most worrisome aspects of pepper processing is the adulteration practiced by some of the more unscrupulous merchants involved in the pepper trade. When such adulterated pepper is exported and the adulteration subsequently detected at the port of entry, it takes not only a heavy toll on the credibility of the exporting country, but, with the recent provisions of the World Trade Organization (WTO), it can lead to international disputes. One of the most common practices of adulterating pepper is mixing it with papaya seeds, which look exactly like pepper (Hartman et al. 1973; Sen and Roy 1974).

7.4 Packaging and Storing

Because it is hygroscopic in nature, pepper can easily absorb ambient moisture, and when such moisture-laden pepper is stored without care, it will lead to the formation of mold, followed by insect infestation because pepper contains a good amount of starch. Thus, great care must be taken when packaging pepper. Mold and insect infestation can lead to loss of aroma, and caking can result, along with hydrolytic rancidity. Whole pepper is generally packed and transported in jute bags (burlap bags) or polyethylene-lined, double burlap bags. Dried pepper having 10–11% moisture can be stored without the mold infestation in jute bags with polyethylene lining or in laminated high-density polyethylene bags (Balasubramanyam et al.

1978). Storehouses should not be damp and should be free of rodent attack, with controlled ventilation and devices for both humidity and temperature control. It is important to fumigate the room before storage. The room walls should be white-washed with slaked lime and should be exclusively used for storing pepper.

Another aspect that affects the processing of pepper is microbiological contamination. Investigations conducted in Malaysia have indicated that pepper berries collected from various farms had a total viable count (TVC) between 10^5 and 10^7/g sample (Apun et al. 1993). The TVC dropped on drying. TVC was found to be less when the produce was mechanically threshed as compared to when threshing with feet. Washing the spikes before threshing and blanching in boiling water minimizes the microbial load.

White pepper is a major pepper product preferred by consumers all over the world. Soaking ripe berries in running water for about 7–10 days is done to make white pepper. The skin of the berries undergoes bacterial degradation, and the softened skin is removed by either decortication or trampling with feet as explained earlier (Pruthi 1993). An investigation done in Sarawak indicated that pepper berries that have a specific gravity >1.12 are the best for conversion to white pepper and are not as good for making ordinary pepper (Anon 1995).

References

Anon (1995) Annual report of the Research Branch, Department of Agriculture, Sarawak, Malaysia

Apun K, Siti Hajijah AS, Yusof IM (1993) Microbiological properties of pepper during processing. In: Ibrahim MY, Bong CFJ, Ipor IP (eds) The pepper industry: problems and prospects. Universiti Pertanian Malaysia, Sarawak, pp 337–341

Balasubramanyam N, Mahadevan B, Anandaswamy B (1978) Packaging and storage studies on ground black pepper (*Piper nigrum* L.) in flexible consumer packages. Indian Spices 15(4):6–11

Esper A, Muhlbauer W (1996) Solar tunnel dryer for fruits. Plant Res Dev 44:61–79

Govindarajan VS (1979) Pepper—chemistry, technology and quality evaluation. CRC Crit Rev Food Sci Nutr 9:1–115

Hartman CP, Divakar NG, Nagaraja Rao UN (1973) A study of identification of papaya seeds in pepper. J Food Sci Technol 10:43–45

Kachru KP, Gupta RK (1993) Drying of spices-status and challenges. In: Proceedings of the National Seminar of Post Harvest Technology of Spices, Trivandrum, India, pp 15–27

Kachru KP, Patil RT, Srivastava PK (1990) Post harvest technology of pepper. Spice India 3 (8):12–16

Krishnamoorthy B, John Zachariah T (1992) Drying black pepper on polyethylene materials. Indian Cocoa Arecanut Spices J 15:75

Lewis YS (1984) Spices and herbs for food industry. Food Trade Press, England, pp 69–77

Madasamy M, Gothandapani L (1995) Spiral separator for cleaning. Spice India 8(11):7–8

Mathew AG (1992) Chemical constituents of pepper. Int Pepper News Bull 16(2):18–21

Pruthi JS (1992) Advances in sun/solar drying and dehydration of pepper (*Piper nigrum* L.). Int Pepper News Bull 16(2):6–17

Pruthi JS (1993) Major spices of India: crop management and post-harvest technology. Indian Council of Agricultural Research, New Delhi, pp 44–105

Purseglove JW, Brown EG, Green CL, Robbins SRJ (1981) Spices, vol 1. Longman, London, pp 1–99

Ramanathan PK, Rao PNS (1974) Equipment needs of spice industry. In: Central Food Technology Research Institute (ed) Proceedings of the symposium on development and prospects of spice industry in India, Mysore, pp 82–84

Sen AR, Roy BR (1974) Adulteration in spices. In: Proceedings of the national symposium on development and prospects of spice industry in India. Central Food Technology Research Institute, Mysore

Variyar PS, Pendharkar MB, Banerjee A, Bandopadhyaya C (1988) Blackening in green pepper berries. Phytochemistry 27:715–717

Chapter 8
An Account of Indonesian Pepper Processing

Abstract The chapter describes, in detail, an account of the Indonesian pepper processing, which is a very important aspect of pepper processing. Oleoresin, another by-product of pepper having odor, flavor, and pungency, is obtained by extracting the berries by using organic solvents. It also details extraction of pepper oil.

Keywords Indonesia · Pepper processing · Oleoresin · Extraction · Packaging · Storing

With Indonesia being a market leader in pepper, it will be informative to review, though briefly, the processing of pepper in that country, more so because the IPC office is located in Jakarta, the capital of the country. In Indonesia, most of the pepper is grown by subsistent farmers. To improve pepper-processing technology, the Research Institute for Spice and Medicinal Plants has carried out research on equipment design meant for pepper processing, which has relevance to on-farm processing. Immediate post-harvest treatment of pepper in Indonesia is quite similar to that practiced in India, but the final cleaning, grading, and packing of the dried pepper are carried out by exporters. This is mainly because the small-scale farmers do not have the infrastructure to carry out the various aspects of processing. Most of the exporters have a complete set of a winnowing and shaking machine (for removal of extraneous materials), a washing machine, a hot air dryer, and a screw-type separation machine for size separation. The entire unit can process around 5 tons of black pepper per hour. Grading is normally done according to requirements of overseas buyer, and double-lined jute bags are used for packing. Packed bags are stored in "godowns" prior to shipment.

In the traditional method of making white pepper, hygienic and quality aspects are the major problems. The berries are usually soaked in rivers or ponds with poor-quality water. Microbial contamination is a major problem, including contamination with waste, and the product acquires a swampy odor that is difficult to eliminate later. In addition, soaking in water for a long time results in the loss of volatile oil and

K. P. Nair, *The Geography of Black Pepper (*Piper nigrum*)*,
https://doi.org/10.1007/978-3-030-52865-2_8

deterioration in aroma and flavor. The quality of white pepper in the whole and ground forms is imparted by appearance, aroma, and pungency. Blanching or steaming fresh green berries for 10–15 min for decortication and then removing the rind in a fruit-pulping machine have been tried (Purseglove et al. 1981). Decortication by soaking in boiling water for 15–25 min has been tried as well, and this led to the white pepper having a better aroma than when prepared by the traditional method (Risfaheri Hidayat 1996). Prolonged soaking in boiling water will lead to loss of volatile oil; it adversely affects the aroma of white pepper because prolonged soaking in boiling water is known to damage the flavor profile of pepper (Pruthi 1992). Soaking in boiling water softens the pepper fruit wall and makes decortication easy.

Pepper oil extraction is an important aspect of pepper processing. It is produced by steam distillation of the pepper berries. Pepper oil does not have piperine, which imparts pungency and so is valued for its aroma in the fragrance industry as well as the flavor industry. Pepper oil is used in high-grade perfumes and the toiletry industry. In the flavor industry, it is primarily used in foodstuffs requiring a high pepper aroma. Volatile oil derived from steam distillation ranges from almost colorless to slightly greenish. Mild in taste with no pungency at all, Indonesia has put much emphasis on pepper oil production and its export.

Oleoresin, another by-product of pepper having odor, flavor, and pungency, is obtained by extracting the berries using organic solvents. The organoleptic properties of oleoresin are determined by its volatile oil and piperine contents, and their abundance primarily depends on the raw material used for extraction. Ground pepper can be extracted with pure organic solvents, such as acetone, ethanol, and dichloroethane. When freshly made, pepper oleoresin is a dark green, viscous heavy liquid with a strong aroma, but, on standing, crystals of piperine appear and so the oleoresin requires mixing before use to ensure uniformity and consistency. The components and the quality of the extract depend on the raw material used, and research in Indonesia indicates that light black berries are better than pinheads for oleoresin, nonvolatile ether extract (NVEE) and piperine (Mapiliandri 1989). Oleoresin offered for sale by some of the principal producers contains 15–20% volatile oil and 35–55% piperine. Purseglove et al. (1981) have reported that as much as 1 kg of oleoresin can be obtained from 8 kg of black pepper. From a review of research on extraction technology, it can be concluded that further research in this area is needed to minimize oil loss and fractionate oil for further blending to produce oil and oleoresin with specific quality characteristics.

References

Mapiliandri I (1989) The extraction of black pepper oleoresin. In: Thesis from Diplome. IV. Ministry of Industry, Indonesia, Ministry of Agriculture (1996). Statistik Perkebunan, Jakarta

Pruthi JS (1992) Advances in sun/solar drying and dehydration of pepper (*Piper nigrum* L.). Int Pepper News Bull 16(2):6–17

Purseglove JW, Brown EG, Green CL, Robbins SRJ (1981) Spices, vol 1. Longman, London, pp 1–99

Risfaheri Hidayat T (1996) Study on decorticating of pepper berries by soaking in boiling water method. Int Pepper News Bull 20(1):17–20

Chapter 9
Industrial Processing of Black Pepper

Abstract The chapter discusses primarily the various aspects of industrial processing of black pepper. After on-farm processing, where drying and grading to the various standards take place, pepper must undergo further industrial processing before the final product is ready for human consumption.

Keywords White pepper · Cryoground pepper · Pepper oil · Oleoresin · Industrial processing

After on-farm processing, where drying and grading to the various standards takes place, pepper must undergo further industrial processing before the final product is ready for human consumption. The dried pepper is passed through mechanical sifters for removal of pinheads, vegetable seeds, fine dust, sand, and the like (all of which are extraneous materials very likely to be present) before winnowing and destoning for removal of stalks, dust, light foreign matter, and stones. Multiple sieve-*cum*-air classifier machines and gravity separators are used for this purpose. The pepper that is separated is then washed in mechanical washers fitted with brushes to remove mold and dust, and this gives a luster to the pepper berries. The product is centrifuged to remove adhering water and then dried in electric or diesel-fired indirect dryers. The dried pepper is subsequently sent through spirals for final cleaning, followed by sterilization either by steam or gamma radiation before being packed. When a pneumatic conveyor-*cum*-dryer-*cum*-grader is used, the entire operation of drying and grading can be done in sequential steps, saving a lot of time.

There are different grades of pepper. Grading is done on the basis of size and shape of the berries. As per the International Organization for Standardization (ISO), the following three grades of pepper exist:

1. Nonprocessed (NP)
2. Semiprocessed (SP)
3. Processed (P)

K. P. Nair, *The Geography of Black Pepper (*Piper nigrum*)*,
https://doi.org/10.1007/978-3-030-52865-2_9

There are permissible limits, set by ISO, for contaminants or light berries, pinheads, broken berries, and the like in the final product and its bulk density according to ISO specification (ISO/DIS 958-1 1996). However, individual countries have their own grades, and these have been detailed earlier. Sterilization is important to ensure good quality and freedom from microbial contaminants. Several methods of sterilization are available, such as hot air/steam sterilization, extrusion, hydrostatic/pressure sterilization, ozone sterilization, compressed carbon dioxide treatment, irradiation, microwave heating, and alcohol treatment.

In hot air/steam sterilization, the product is preheated to 50–55 °C and subsequently sterilized by a combination of indirect heating and direct steam injection. The process substantially reduces microbial contaminants while not adversely affecting the flavor, aroma, or oil content (Schneider 1993). Yet another process of steam sterilization involves exposure to high pressures and temperatures for a predetermined time period in a series of chambers. The material to be sterilized is placed in the first chamber; high pressure is exerted and later transferred to subsequent chambers, following a similar procedure for specific time periods; and finally, the sterilized material is depressurized in a chamber (Dudek 1996). There is a modified steam sterilization process that consists of a steam jacketed pressurized vessel with three temperature envelopes, namely, 50, 100, or 200 °C, for different products. Because the system is pressurized, there will be no loss of oil or moisture, and the appearance of the product remains unchanged. Steam, compressed air, and nitrogen gas may be used depending on the end product desired (Darrington 1991). A continuous steam sterilization process developed involves subjecting the pepper to a rapid flow of superheated steam during a predetermined time period, followed by drying, rehumidification, and packaging. Both enzyme and microbial activities are significantly reduced to low levels, and no loss of flavor or aroma or oil was noted (Uijil 1992). Leife (1992) describes a system of steam sterilization where pepper is subjected to rapid pulses of steam, the steam condenses on the surface of the berries, and contaminants are removed. Steam sterilization is the ideal method in situations where, due to national safety restrictions, both chemical and irradiation methods are prohibited.

Yet another method of sterilization is gamma irradiation. This method is particularly useful against insect infestation. Because the process is a cold treatment, loss of volatile oil, flavor, and aroma are practically nil (Nair 1993). A low dose of 1 kGy (kiloGray) is sufficiently effective against insect infestation, whereas a tenfold increase is needed to control microbial infestation. The process does not lead to accumulation of any harmful residues, unlike fumigation. Depending on the source geometry and the conveyor system used, the size and shape of the containers will be decided. Codex General Standards for irradiated food and recommendations of the International Code of Practices for the operation of irradiation facilities give the guidelines for irradiation procedures. It is important that Codex Standards are followed to facilitate international trade. Pepper subjected to a radiation level of up to 10 kGy does not lose its aroma and flavor, and there are no significant changes in the oil or piperine contents. When the produce is heavily contaminated, irradiation up to 30 kGy is needed. Countries such as USA, France, Austria, and Germany have approved this irradiation procedure for sterilization, and a commercial application of the technology is followed. More than 95% of the gamma irradiation facilities operating in the world use cobalt-60. Gamma irradiation facilities are capital-intensive, and the unit cost of processing is highly sensitive to the scale of operation. To minimize operational costs, facilities need to operate round the clock at the maximum rated throughput that the facility is designed for. A gamma irradiator is simple to operate and maintain. Chemiluminescence, thermoluminescence, and free radial dosimetry are employed as routine methods for the detection of irradiated pepper. Other methods in use are colored indicators on the containers that are to be irradiated, viscosity measurements of gelatinized pepper, and intensity measurements of reflected signals of the electron spin resonance (ESR) spectrum.

Two other methods of sterilization use chemicals and microwaves. Fumigation with ethylene dibromide to disinfect insect contamination followed by fumigation with ethylene oxide or propylene oxide for eliminating microbial contamination is the most commonly used chemical sterilization process. Moisture content in the berries, temperature, time of contact, and concentration of the fumigating gas decide the effectiveness of the sterilization process. Fumigation results in the reduction of both volatile and nonvolatile oil contents. The main problem with the use of these fumigants is their explosive, toxic, and irritant nature, and as such, they are unsafe when used without adequate safety measures for the workers who are involved. Additionally, harmful residues are left behind following fumigation. On account of these problems, many countries have banned the use of these fumigants. The main advantage with microwave sterilization is that the microbial count is drastically reduced. Microwave heating is done at 2450 MHz and high-frequency heating at 27.12 MHz. With an increase in moisture in the samples, microbial efficiency is increased, but there are losses of both volatile and nonvolatile oils. Both chemical and microwave fumigation are comparable in their efficiency.

A number of value-added pepper products of industrial origin exist. The first is dehydrated green pepper. This product is prepared by industrial processing from immature green pepper fruits of appropriate varieties. The fruit should be uniformly sized with the pungency, flavor, and color of green pepper. The fruits are blanched in boiling water for a few minutes; the water is drained; the fruits are cooled and soaked in SO_2 solution to fix the green color, followed by drying in a cabinet dryer at 50 °C. Upon rehydration, the product will reconstitute to a good-quality product possessing the characteristic pungent spicy taste, color, and flavor of green pepper; reconstitution is achieved when 1 part by mass of dehydrated green pepper is cooked for 20 min in the presence of 10 parts by mass of a 1% concentration of NaCl solution. To conform to international standards (ISO/DIS 10621 1996), the product should have a moisture content of less than 8%. The use of SO_2 is a health hazard, and efforts are under way at different research institutions to find an alternative (Sankarikutty et al. 1994).

The next in line among the value-added products is canned green pepper. Of late, there has been a surging demand for canned green pepper. After removing the spikes, the pepper fruits are washed in running water and kept soaked in water with a chlorine concentration of 20 ppm for 1 h. They are then covered with a 2% hot brine of 0.2% citric acid warmed to 80 °C, sealed airtight, and kept in boiling water for 20 min. After this, they are immediately cooled in a stream of running cold water. A better color is obtained using acetic acid instead of citric acid, but as per international standards, only citric acid is permitted, not exceeding 0.6% by mass of the packing medium, and the covering brine must be 1–2% by mass of edible common salt. For making canned pepper, berries harvested a month prior to the actual date of maturity are the best.

Bottled green pepper, dry packed green pepper, and freeze-dried pepper are among the other directly edible pepper products. Bottled green pepper is made by first despiking fresh green pepper fruits of uniform size and maturity, immediately after harvest, followed by cleaning, washing, and steeping in 20% brine solution containing citric acid. This is then allowed to cure for 3–4 weeks. The liquid is then drained off, and fresh brine of 16% concentration together with 100 ppm SO_2 and 0.2% citric acid is added. The resulting product is stored in containers kept away from direct sunlight. As per international standards (ISO/DIS 11162 1996), the product will have the characteristic odor and flavor of fresh green pepper, the color varying from pale green to green. Dry packed green pepper is produced in the same way as bottled green pepper, with the only exception being that the liquid at the final stage is drained off and packing is done in flexible pouches. Dry packed green pepper, which has similar qualities to that of canned or bottled green pepper, can replace the latter. Freeze-dried pepper retains the original color and shape of green pepper and fetches a premium price in the market. Its processing is a manufacturer's secret.

9.1 White Pepper

White pepper is produced from fully ripened fruit after removal of the outer pericarp either after or before drying. White pepper is preferred for use in food products, such as colored sauces, salad dressing, soups, and mayonnaise, where dark-colored (black) pepper is undesirable. In some European countries, it is white pepper that is in common use. It is made by any of the following methods: (1) steeping and retting in water, (2) steaming, (3) boiling, (4) chemical treatment, or (5) the simple decorticating process of ripened fresh or dry berries.

The water steeping and retting (or soaking) process is started when one or two fruits in the spike start yellowing when the crop is harvested and readied for treatment. After harvest, the produce is threshed and heaped in tanks through which water is allowed to run for 7–10 days. Lightweight pinheads and light berries accumulate on the surface and are removed, and the remaining mass is rolled over at least three times a day during the retting stage. On the 11th day, the outer skin of the berries is removed by gentle rubbing, and the deskinned fruits (the pepper seeds) are transferred to another tank containing a bleaching solution. The produce is left to stand in the bleaching solution for 2 days, after which the solution is drained and the seeds washed and sun dried. The process involves the steaming or boiling of the mature green fruits for 10–15 min when the outer skin of the fruits gets softened, which is then removed by passing through a pulping machine. The deskinned fruits (pepper seeds) are washed and treated with SO_2 or bleaching powder solution and subsequently washed in water and dried in the sun. The skin of the berries recovered in this process can later be used for the recovery of pepper oil by steam distillation, though it may not be economical. The process has been developed by the Central Food Technological Research Institute in Mysore, India.

There are a number of chemical processes that can be used to prepare white pepper by treating the berries with acids or alkalis. As of now, no commercially viable production unit exists. Manilal and Gopinathan (1995) have described a method by which the skin of both dried and fresh pepper is removed by microbial decortication. By using decorticating machines, whole dried black pepper fruits can be processed to white pepper. Because loss due to breakage will be high, production of white pepper by this method will be expensive. The characteristic flavor and aroma also could be missing in the final product, and as such, this method of making white pepper is not recommended.

Among all the processes described herein, the traditional retting process is the most popular, and white pepper obtained by this method is preferred by consumers. According to international standards (ISO/DIS 959-2 1996), the product should have a maximum of only 1% permissible extraneous matter, 4% of broken berries, and 15% of black berries. The minimum bulk density should be 600 g^{-1}. On a comparative basis, while 100 kg of mature green pepper will yield 33 kg dry pepper, in the case of white pepper, it will only be 25 kg. From this, it can be inferred that the cost of white pepper should be at least 35% above that of black pepper. The total world demand for white pepper is around 38,000 metric tons per annum, which is a quarter of the worldwide black pepper production.

There is yet another value-added pepper product known as ground pepper. It comprises both white and black pepper. Ground pepper is obtained by grinding the normal pepper without adding any external material. Pepper has to be ground to a specific size. For grinding, hammer mills with copper-tipped hammers are preferred to silica-tipped plate mills because ground pepper obtained from the latter can contain silica in excess of the permissible limit. The ground material, after sieving, is packed, and any material above the specified size is sent back to the mill for further grinding and size reduction.

When pepper is processed in modern spice grinding units, it is first cleaned of any extraneous matter and then passed through a magnetic separator for the removal of metal particles. After this, it is passed through a vibrating screen for further removal of extraneous matter and then sent to the hammer mill. The ground product from the hammer mill is fed into a cyclone separator to recover pepper powder. It is further screened, using appropriate sieves, and the overflow is recycled. Bag filters are used after the cyclone separators to prevent outside escape of fine particles. The ground pepper is packed into airtight containers prior to shipping, and the packages are made to specifications provided by the customer on size, pungency, moisture content, ash content, crude fiber, volatile oil, piperine, starch content, and the like. White pepper is processed in the same manner as that described earlier, except instead of black pepper, the starting material is white pepper. The ground pepper has a characteristic flavor—very aromatic and slightly sharp—and must be free of any extraneous odors and flavors, including moldy and rancid odors. The international standards for production of ground white pepper conform to ISO/DIS 959-2 (1996). White pepper can also be made from black pepper, using selective grinding followed by sieving. The skin of the berries is a by-product and contains essential oil and oleoresin, both of which can subsequently be extracted.

9.2 Cryoground Pepper

To cut down oxidative losses of essential oils and to assist in the fine grinding of pepper by making the raw product brittle at low temperatures, a new technique—known as *cryogrinding*—is used at temperatures very much lower than the normal 100 °C. The product disperses more uniformly in spice formulations. The process involves injecting liquid nitrogen into the grinding zone. A temperature controller maintains the desired product temperature by suitably adjusting the liquid nitrogen flow rate. The exhausted gas is recirculated for the precooling process.

9.3 Pepper Oil and Oleoresin

Pepper oil and oleoresin are the two most important ingredients of pepper and have great economic value. The first imparts the aroma to the pepper, while the second provides the pungency and flavor. The essential oil can be recovered by steam or water distillation. The essential oil contains mainly a mixture of terpenic hydrocarbons and their oxygenated compounds, having a boiling point in the range of 80–200 °C. Depending on the pepper cultivar, the agroclimatic situations in which the crop is grown, the management aspects, and the grades, the content of the oil varies. The pinheads contain the most sesquiterpenes. In order to meet different flavor requirements, it is possible to suitably blend oils from different grades of pepper.

In industrial processing to recover essential oils, the product is first flaked using roller mills and then steam distilled in stainless steel extractors. The product to be distilled is heaped into the distillation unit and then compacted near the walls to prevent any steam escape. Through the bottom of the still, dry steam is passed. The still also gets heated through the jacket provided for this purpose. When the steam comes into contact with the pepper, the temperature is raised, and the oil present in the oil cells of the berries vaporizes, rising along with the steam through the still. It is then condensed, and because the oil is lighter than water, it floats on the surface of the condensing water. Using an oil or water separator, the oil is then recovered. The oil recovered by steam distillation is allowed to remain on anhydrous Na_2SO_4 to exclude traces of moisture. When exposed to direct sunlight and air flow, the hydrocarbons and the sesquiterpenes present in the oil undergo oxidative changes, especially during long storage. Hence, it is always desirable to store the material in airtight containers, especially because the end use occurs after a long time as a result of export via shipping to different countries.

Oleoresin is usually obtained by extraction of ground pepper with acetone, ethanol, ethylene dichloride, and ethyl acetate. The main advantage of oleoresin is that it is uniform in composition and strength. Contaminants such as mold and fungus are absent in oleoresin and can be directly added to foodstuffs after adjusting the level of concentration. Oleoresin comes in oil-soluble, water-dispersible, and dry forms. Currently, oleoresin is recovered either by a single-stage or by a two-stage process. The process involves size reduction of the ground pepper prior to solvent extraction done in stainless steel extractors.

In the single-stage process, the oil is recovered (along with the resins) by solvent extraction, whereas in the two-stage process, the ground pepper is first subjected to steam distillation for the recovery of the essential oil. The composition of the oil and the oleoresin content obtained in both the processes differ slightly from each other. Due to the moisture present in the wet cake obtained after steam distillation, there is a likelihood that the oleoresin yield obtained in the two-stage process will be less compared with that obtained in the single-stage process. Drying prior to solvent extraction can prevent such loss in oleoresin yield. In the case of the single-stage process, pepper is flaked to 1.0–1.5 mm thick and is packed in stainless steel extractors for extraction with organic solvent. Normally, a solid-to-solvent ratio of 1:3 is maintained at an extraction temperature of 55–60 °C. The solvent is continuously recirculated to ensure efficient solid-to-solvent contact; after 3 h of extraction, the miscella is filtered and evaporated. The solids are further contacted with lean solvents from other extractors, and the entire extraction process is completed after six or seven stages of extraction. The miscella sent from the extractors are evaporated in a shell and tube evaporator at temperatures not exceeding 80 °C in a vacuum of 250 mmHg. As the temperature starts to rise and no further solvent is recovered from the concentrated miscella, it is pumped into a high-vacuum stripper. Final desolventization is done at a vacuum of less than 20 mmHg; at no stage of the operation is the temperature allowed to rise above 100 °C. Initially, during vacuum distillation, the condensate recovered will mostly be the organic solvent; toward the end of the distillation, the essential oil will start coming out along with the organic solvent. At this stage, some entrainer, such as alcohol or the monoterpene fraction of the essential oil, is added to the desolventizer. This is done to remove the final traces of the organic solvent as well as to supplement the monoterpene fraction of the pepper oil that might have been lost during the final solventization. The product is pumped into storage tanks after confirming that the residual solvent levels are within limits. Suitable blending is used to meet the customer requirements in terms of both oil and piperine contents. Some customers specify the homogeneity of oleoresin through a homogenizer, which may be a colloid mill or sand mill. The oleoresin, which is dark in color, is bleached using activated carbon to obtain decolorized oleoresin. All over the world, oleoresin extraction is done mostly in batch extractors. Investigations conducted at the

Regional Research Laboratory, Trivandrum, in Kerala State, India, showed that extraction can be done in ambient conditions and at low residence time (Sreekumar et al. 1993). Present-day extraction, which takes more than 18 h for the processing of one batch, can be dispensed with by employing continuous countercurrent extraction. Simultaneously, the quantity of solvent needed for extraction and its loss in the extraction process can be substantially reduced.

Research in the laboratory has shown that extraction by enzymatic breakdown of spice cell walls of the dried pepper can be accomplished. In this process, the product is mixed with water, adjusting the pH by the addition of citric acid, treated with enzymes, either individually or in combination, which is followed by centrifugation. The enzymes used are those that are commercially available, such as cellulase pectinase, hemicellulase, and liquefaction enzyme preparations. Extracts having good sensory and compositional properties were obtained with some of the enzyme combinations. Optimum results were obtained using a combination of cellulase and pectinase preparation. The addition of hemicellulase did not result in any improvement of flavor. This solvent-free extraction procedure is an alternate route for the recovery of flavor compounds that is not commercially possible at the moment. Another solvent-free extraction procedure that has shown much promise is supercritical carbon dioxide extraction. Low-energy consumption, high purity of resultant extracts, environmental acceptability, and possible fractional separation of the components are some of the major advantages of this process.

The current demand for piperine is on the increase. Piperine can be produced from the oleoresin in concentrated form by centrifuging the oleoresin in a basket centrifuge. Part of the oil, along with some resin, can be collected after centrifugation, and the centrifuged cake, which contains as much as 60% piperine, is obtained. By washing the centrifuged cake with pepper oil and further centrifuging, the piperine concentration can be additionally enhanced.

A number of secondary products have been developed from oleoresin to improve solubility in food substances that are marked under various trade names by the manufacturers, with their flavor strength indicated on the respective label. Standardized seasonings, which are able to withstand almost all processing conditions, are some of these secondary products. Emulsions, solubilized spices, dry soluble spices, encapsulated spices, heat-resistant spices, and fat-based spices are some of the additional secondary products. Emulsions are liquid seasonings that are prepared by emulsifying blended pepper oil or oleoresins with gum acacia or other permitted emulsifying agents. A stabilizer is added to check creaming, and the products have only limited shelf life. Solubilized pepper is blended pepper oil and oleoresin that are mixed with one of the polysorbate esters in such a concentration to give a clear solution when mixed with fresh water. Dry soluble pepper is prepared by dispersing the standardized oleoresin onto an edible carrier-like salt, such as dextrose or rusk, in order to provide a product that has a flavor strength equal to that of good- or average-quality ground pepper.

In addition to the industrially produced, value-added by-products of pepper previously described, there are a few others, such as heat-resistant pepper, fat-based pepper, extruded spices, and microencapsulated flavors. Of these, microencapsulation is the most important. It is a process by which the flavor material is encapsulated in a solid matrix and is ready for release as and when required. Encapsulation can be achieved by a number of techniques, such as spray-drying, coacervation, and polymerization. Of these, spray-drying is the most popular. The process involves homogenization of the oil/water mixture in the presence of the wall material and later removal of water under controlled conditions in a spray dryer. The advantage of the spray-drying process over others is that the product, though in contact with existing gas at a higher temperature, will never reach this high temperature within the short residence time. The oil/water emulsion is atomized in the spray dryer. Commonly used wall materials for encapsulation are selected from among vegetable gums, starches, dextrins, proteins, sugars, and cellulose esters. The wall material is selected so as to meet, as closely as possible, the desired properties, such as low viscosity at a high solid state; ability to emulsify or disperse the active material; nonreactivity with the material to be encapsulated, both during processing and in prolonged storage; uniform film-forming; ready availability; and the like. Investigations indicate that the addition of surfactants during the process of encapsulation prior to spray-drying will reduce the oil loss (Anon 1989). Another process for microencapsulation is the CR-100 process (Anon 1995; Findlay-Wilson 1995; Mos 1995), which overcomes the limitations of the spray-drying process such as reduction in flavor quality and yield.

Heat-resistant pepper is a double-encapsulated product in which the capsules are rendered water insoluble by a suitable coating and the contained flavor is released only at high temperatures, such as in the case of baking. Fat-based pepper is a blend of pepper oil and/or oleoresin in a liquid edible oil or hydrogenate fat base formulated for use in products such as mayonnaise. Extruded spices are those sterilized, ground, and encapsulated in a single step. The process involves a combination of pressure changes, temperature shock, and shear. The process, accomplished in a twin-screw extruder, helps retain color and flavor in the original form because of the very short processing time. When fresh spice is fed into the twin-screw extruder, the starchy materials become gelatinized and form an encapsulated product. This process substantially reduces contamination by bacteria, mold, and yeast. The product emerges from the extruder as a "spice rope," which is then cut into pellets. Lucas Ingredients, Britain, markets the products under the brand name Master Spice (Scott 1992).

References

Anon (1989) Technology transfer document of microencapsulated spice flavors. In: Regional Research Laboratory, Council of Scientific and Industrial Research, New Delhi/Trivandrum, pp 9–10

Anon (1995) Annual report of the Research Branch, Department of Agriculture, Sarawak, Malaysia

Darrington H (1991) Herbs and spices. Food Manuf 66(11):21–23

Dudek DH (1996) Sterilization method and apparatus for spices and herbs, US Patent US 5523053 [US 260068 (940615)]

Findlay-Wilson IZFL (1995) Microencapsulated flavours for a great taste. Int Zeitsch Lebensmittel—Techn Market Verpack Anal 46(10):76–78
Leife A (1992) Sterilization of spices with pulsed steam. Livsmedelsteknik 34(1/2):24–27
Manilal VB, Gopinathan KM (1995) Annual report of Regional Research Laboratory, Council of Scientific and Industrial Research, Trivandrum, Kerala State, India
Mos AM (1995) Microencapsulation enhances flavourings and ingredients. Voedingsmiddelen-technologie 28(23):64–65
Nair PM (1993) Spice irradiation—present and future scenario. In: Narayanan CS, Sankarikutty B, Nirmala Menon A, Ravindran PN, Sasikumar B (eds) Post harvest technology of spices, Indian Society of Spices. Indian Institute of Spices Research, Calicut, pp 28–33
Sankarikutty B, Padmakumari KP, Howa, U, Narayanan CS (1994). Annual report of Regional Research Laboratory, Council of Scientific and Industrial Research, Trivandrum, Kerala State, India, pp 7–8
Schneider B (1993) Steam sterilization of spices. Fleischwirtschaft 73(6):646–649
Scott R (1992) Master spice: extruded low count spices. In: Proceedings of the New Technologies Symposium Bristol, UK, pp 6–8
Sreekumar MM, Sankarikutty B, Nirmala Menon A, Padmakumari KP, Narayanan CS (1993) Annual report of Regional Research Laboratory, Council of Scientific and Industrial Research, Trivandrum, Kerala State, India, pp 5–6
Uijil C dH (1992) New continuous steaming method for herbs and spices. Voedings-middelentechnologie 25(6):40–43

Chapter 10
The Future of the Global Pepper Economy

Abstract The chapter discusses in great detail the future of Indian pepper economy vis-à-vis the global pepper economy. The global pepper economy has different facets: (1) the supply side, (2) the demand side, (3) price projections (both supply and demand included simultaneously), and (4) pricing and market equilibrium. These facets are discussed with regard to many pepper-producing countries.

Keywords India · Production · Marketing · Pepper market · Pepper economy

The global pepper economy has different facets: (1) supply side, (2) demand side, (3) price projections (both supply and demand included simultaneously), and (4) pricing and market equilibrium. Bade and Smit (1992) have devised precise models to study these facets. This review does not provide an elaborate discussion of these models, but the essential conclusions stemming from them will be discussed in this section. The various models are all mathematical in nature.

10.1 The Supply Side of the Pepper Economy

The supply side of pepper is approached on a national basis, examining the area, production, and exports from each country. Normal production value is obtained by multiplying the number of pepper vines by the average yield of a vine in a country. Although measurement by vines must be considered much better, a similar line of reasoning can be applied with area and average yield per hectare. The major drawback of this method in comparison with using vines is that it adds to a source of variation. Not only the per vine yield fluctuates, but also the number of vines per hectare. In India, this is a very significant variation, but in other countries—for example, in Thailand— the number of vines per hectare is almost the same all over the country. If the intensity of cultivation is relatively constant and there is no significant change in cultivation patterns, it is possible to estimate a nationwide

K. P. Nair, *The Geography of Black Pepper (*Piper nigrum*)*,
https://doi.org/10.1007/978-3-030-52865-2_10

normal average yield per hectare. In case the average number of vines per hectare cannot be expected to be constant, an assumption is needed about the change. In case the intensity of cultivation increases or a superior variety is introduced, it can then be assumed that there will be an upward trend in the normal yield per hectare some 2 years after the start of the intensification when the planted vines become productive. If normal yield can be used as a basis for forecasting, the actual yield will deviate from it because of the influences of weather, amount of fertilizer applied, and level of maintenance. Unfortunately, these effects are not confined to only 1 year. When a year is very wet, the crop will be small, but most importantly, there will be also the incidence of diseases such as foot rot. Invariably, such diseases will determine the yield in the succeeding year. Also, the effects of neglect or exceptionally good maintenance will spread over more than 1 year.

The preceding discussion illustrates that even when data on productive vines and average yield are available and quite reliable, there would be sufficient scope for simulation and expert interpretation. However, global information is very expensive to procure, and gathering information on agriculture can be even more expensive and time-consuming. This is not a priority of the developing countries. Planning depends largely on information, and a model cannot compensate for the lack of quality data. Inasmuch as quality of information is concerned, it can only interpret and detect inconsistencies. Thus, in conclusion, it can be said that the modeling of production and supply presented is only a preliminary step on the road to a more sophisticated model analysis based on superior data. In the following paragraphs, a brief review of the pepper economy of important pepper-producing countries, such as India, Brazil, Thailand, Malaysia, and Sri Lanka, is given.

10.1.1 India

It is a very important observation that the pepper-growing area (also known as the area "under pepper") in India would be enough to meet the entire global demand if the average yields were only a third of what is obtained in Sarawak (Malaysia). This points to very important lacunae in data collection, precision, and reporting. Except in large pepper plantations in the states of Karnataka and Northern Kerala, pepper is still grown in homestead gardens as an intercrop between coffee and cardamom and is quite often trailed using areca nut and coconut trees as standards. Up to 1986, a survey was conducted in Kerala—by field extension agents in randomly chosen parts of Kerala—in such a way that within 5 years, every part was visited once. The total area under pepper from the population was then multiplied by the inverse of the sample area, divided by the state area ratio. The goal was to estimate area on the basis of 560 vines ha^{-1}. The method was applied, asking the same people, to get production estimates. From 1987 onward, the Department of Economics and Statistics of the state has used a more sophisticated system, especially to estimate production. Though the method used is rather elementary, other countries probably do not have any better system. A similar equation as that used in India is also used in Brazil.

On the market side, it is seen that an increase in price is an incentive for enhanced acreage as more farmers resort to the use of fertilizers and adopt better maintenance practices followed by more picking rounds. The previous year's price is the trigger for these added incentives. Farmers also keep pepper in stock, hoping to sell it for a better price in the following year if a price drop is encountered in the current season. Alternatively, when prices escalate and farmers release their stocks, it will seem as though production has gone up, which is not the case. Indian exports depend largely on the current season's crop. The other factor is the influence of price changes on stocks of traders. When there is a sudden increase in price, the total amount of exports may even exceed production. The year 1985 was exceptional. Again, 1991 and 1992 were also exceptional on account of the restructuring of the Indian economy when globalization and liberalization processes set in.

10.1.2 Brazil

In Brazil, there are no time-series data available on area by cultivation system, although there are different cultivation systems in the country. To be sure, it is still totally unclear how the Brazilian Pepper Exporters Association gathers the data presented at the IPC meetings. Apparently, data on area and production do indicate that pepper yields have gone up over the years in Brazil. But this is not the case. If production is divided by area, it would appear that there is a decreasing trend over time. In fact, between 1978 and 1979 yields plummeted—despite area expansion—suggesting that in those years, data on area did not include area with immature vines. After 1979, the correlation between area and production was strong, although 1984 was an extraordinary year that had a much higher yield than could be explained. This may have been due to positive price correlation.

Pepper consumption in Brazil is negligible compared with production and presumably kept-out-of-production statistics, and because there are some other small pepper-producing regions outside of Para state, it can be expected that almost all pepper produced will be exported. However, there is a rather constant amount of about 1500 tons that is apparently not exported. Price is the most important variable in the current productive area under pepper.

10.1.3 Indonesia

Indonesian data on aggregate area under pepper are rather scanty. The official records claim that the total area did not change from 1983 until 1987. However, records for Lampung and Bangka show substantial changes over these years. The time-series data available on the areas of Lampung and Bangka are still too minimal.

Further, there is hardly any information on area in Kalimantan. For future modeling research, regional disaggregation of supply of Indonesia is important, especially because of the special status of Bangka, which produces only white pepper. Prices have played only a minor role in area expansion. Clearing of new land for pepper production will play a predominant role in pepper production, especially in Kalimantan and Sulawesi. For Indonesia, the influences of price on stocks are far less important than on maintenance, a similar conclusion to that for India. The year 1993 was a bad year because of a serious setback in production and thus on exports, reflecting a serious situation for white pepper in Bangka.

10.1.4 Thailand

Thailand is a relative newcomer to pepper production. Until 1986, area under pepper was quite steady; thereafter, it expanded, but once again dropped between 1991 and 1993 when low prices ruled the market. At this time, the government encouraged farmers to diversify from pepper. Changes in prices were influenced by prices prevailing in the previous year, while area was held steady. Productivity per hectare is very high, and the production is influenced by the previous year's prices. Over recent years, production has been on the decline. The country has a high domestic consumption, which is around 5000 tons per year. Exports take a fixed share of production, which is influenced by the previous year's shortage or surplus.

10.1.5 Malaysia

Malaysian data on area and production of pepper, of which nearly 98% refers to that of Sarawak, are rather unreliable. Foot rot strongly influences yield. Data on production are based on exports; pepper farmers from Sarawak speculate on the pepper market, so export is not a reliable index to show production trends because the farmers who speculate in the pepper trade are rich. Hence, large differences between production and exports are bound to arise. From 1989 onward, there is a decline in area arising from shortage of labor and its high cost, which did not allow for profitable production to the extent as was the case in the past. The price of the previous 2 years greatly influenced production per hectare, though this influence is through the planting of new vines.

10.1.6 Sri Lanka

A traditional pepper producer, there is only a slight increase in area over the years in Sri Lanka. Productivity is low, and despite an increase in area, production did not correspondingly increase, which is rather surprising. The production pattern in 1993—a year that showed an incredible increase in production and exports—is just the opposite that of Indonesia. Export in 1990 was exceptionally high, exceeding production. Export can be explained on the basis of production, the change in production, and domestic prices.

10.1.7 Other Countries

In addition to the countries specifically discussed, countries such as Madagascar and those on the African continent (Benin, Cameroon, Congo, Gabon, Ethiopia, Nigeria, Zambia, Zimbabwe) also produce pepper. Madagascar, with a production of about 2500 tons, leads the pack. Because long-term time-series data and information on area and production are needed, as well as information on internal markets, no tangible conclusions can be drawn from the existing information.

10.2 The Demand Side of the Pepper Economy

To obtain a clear analysis of the demand side of the pepper economy, precise data on the end use of pepper are needed. To date, precise data on differentiation of pepper use in food industries, institutional catering, and household consumption are unavailable, and as such, only rough estimates can be made. Further, a distinction between black and white pepper is needed. The following countrywise groupings will provide an insight into the demand side of the pepper economy.

10.2.1 European Union

As a shift in the food consumption pattern takes place—for example, consumption of more prefabricated food, fast food, and the like, the demand scene will fluctuate. The countries of the European Union are price conscious as far as stock formation is concerned. Also, the level of stocks at the end of the previous year has a reasonably significant influence.

10.2.2 Rest of Western Europe

Starting from 1998, Switzerland has import data on pure pepper; these figures previously included pimento and capsicum. Price changes influence stock formation.

10.2.3 Eastern Europe and CIS

The economic problems that overtook Eastern Europe and the Commonwealth of Independent States (CIS) have had a very negative effect on import since 1990. Price changes influence stock formation.

10.2.4 North America (USA and Canada)

There is a negative relationship between price levels and starting stocks in the USA and Canada. When prices are low, traders will keep more stock, expecting prices to increase again sometime in the near future; if stocks are high, traders tend to sell.

10.2.5 Japan

Demand for pepper in Japan is pegged to the gross domestic product (GDP) in more than one way. A shift toward outdoor, ready-to-eat food, in particular Western food, goes along with a rising GDP. Changes in stocks paralleled trends in North America.

10.2.6 Australia and New Zealand

Stock levels are related to consumption in Australia and New Zealand, and no price influence is found on the demand side.

10.2.7 Latin America

Per capita import of pepper did not show any increase in Latin America. Between 1976 and 1980 and 1985 and 1988, significantly lower imports took place, and stock formation depended on prices and lagged stocks.

10.2.8 Asia and Pacific Countries, Singapore, Australia, and New Zealand (Excluding China)

Stock formation depended on prices and lagged stocks in the Asia Pacific area.

10.2.9 China

China was a small importer of pepper, and the import spurted in the latter part of the 1970s. Of late, imports have dwindled.

10.2.10 Middle East and North Africa

The Middle East and North Africa have a rather irregular import pattern. Stock formation was influenced by price changes.

10.2.11 Rest of Africa

Per capita import of pepper did not show any increase in the rest of Africa, as was the case in Latin America. Since 1990, imports surged. Stock formation was influenced by price changes.

10.3 Prices and World Market

As far as the world market is concerned, Singapore holds a special place for pepper import and its commercial consequences. Of the total pepper exported from Malaysia, 69% moves via Singapore. The function of Singapore as an entrepôt is accounted for by the variable total exports of producing countries less estimated world consumption, which is merely an estimate of the change of stocks outside producing countries. Singapore is expected to import some of these stocks while keeping the major part of these stocks as carryover. Some part of pepper imports is consumed, but no statistics are available on pepper consumption in Singapore. It is estimated that consumption is approximately 1% of imports. The price pattern indicates that pepper traders are presumably more interested in trade if prices are high, which does not seem unrealistic because margins will definitely be correlated with price peaks.

10.4 The Pepper Price Outlook by 2020

Perhaps the most important aspect of the price structure is the impact of the size of the Indian supply responses to price fluctuation. This aspect has not been sufficiently captured. For all countries, the investment side is not yet represented adequately for long-term analysis. Stock formation at various levels needs further work as well. Price cycles of around 7 years will continue in the future, showing that the 10-year cycles in income have very little impact. There will not be an increase in average real prices. Obviously, with inflation around 4% per year, nominal prices do increase on average.

10.5 The Pepper Supply Outlook by 2020

Projections indicate that production and exports are expected to decline in Brazil, Malaysia, and Thailand because of the increase in labor costs. (India and Indonesia, however, show strong growth.) Brazil shows a modest decline on average, and the same conclusion can be drawn for Malaysia and Thailand. These countries are becoming too rich to grow labor-intensive agricultural crops such as pepper. Sri Lanka is expected to level off on account of limited area. Exports from Vietnam and China, especially the former, are expected to grow rapidly, reaching an aggregate of more than 60,000 tons. A large quantity of pepper is imported into India from Vietnam, often clandestinely, in the garb of value addition. Because Indian pepper fetches a premium in the world market, unscrupulous traders mix the low-quality and cheap Vietnamese pepper with Malabar pepper and reexport it as Malabar pepper to the USA and Europe. On detection by the consumers, this adversely affects Indian pepper in the world market, pricewise, and also affects future imports.

The gap left behind as relatively rich countries such as Brazil, Malaysia, and Thailand decrease their production and exports is filled by Vietnam and China—but not by traditional major producers such as India and Indonesia. Exports from Madagascar and other African countries are expected to be steady. This inevitably leads to the important conclusion that almost all pepper exports will originate in Asia.

10.6 The Pepper Demand Outlook by 2020

On the consumption side, North America, Japan, and the European Union will show a steady growth. Developments in Eastern Europe will affect the consumption pattern, so will the increased consumption in the Pacific countries and other upcoming Asian regions, such as the Philippines, Korea, and Cambodia. Another major consumption area is the Middle East and North Africa.

10.7 Countrywise Economic Growth Affecting Production and Consumption

It is interesting to analyze the GDP/growth rates that affect pepper production and consumption. The details of the estimates are presented by Burger and Smit (1997). Historical developments have affected consumption. For instance, the late 1980s were years of political turmoil for Eastern Europe and the CIS, and this adversely affected growth rates and consumption. On the other hand, countries in eastern, far eastern, and southeastern Asia have enjoyed very buoyant growth rates, which positively affected consumption. However, such very high growth rates cannot be maintained for decades, as has been clearly shown by other countries with high growth rates in the past.

In the long run, the USA will average a growth rate of just below 2% per annum, while Japan, despite its recovery, may not reach the US growth rate after its period of high growth up to 1990. In Western Europe, growth rate will hover around 1–3%. Eastern Europe, CIS, and former Yugoslavia are expected to recover but, on average, at rather moderate levels of economic growth. Growth rates in Latin America will be somewhat higher than in Europe. In south Asia, India is maintaining a growth rate around 5%, though expectations were higher. Pakistan has a somewhat higher rate. In Southeast Asia, the Philippines will show a moderate growth, while in other countries, though showing higher rates, a declining trend has been observed. The same applies to China. Vietnam will grow around 4%. And, with the exception of Nigeria, Africa on the whole is showing only very moderate growth rates.

The preceding growth scenario indicates that the potential for pepper export from producing countries to the USA and Europe remains bright in the long run. This is primarily because the processed and fast-food cultures will unfold further in the years to come. Southeast Asian countries are also potential markets. There is not much scope for export into Africa. Latin America, which will have a higher growth rate, is a potential market, but countries such as Brazil are already important pepper-producing countries, and as such, export from other pepper-producing countries has only limited scope for export to Latin America.

References

Bade J, Smit HP (1992) An annual model of the world pepper economy. Seminar on commodity analysis, Jakarta, Indonesia. ESI-VU, Amsterdam, p 63

Burger K, Smit HP (1997) The natural rubber market. Woodhead Publishing Limited, Cambridge, p 354

Chapter 11
Pepper Economy in India

Abstract The chapter discusses all aspects of Indian pepper economy as India is the most important producer of black pepper in the world. In India, pepper is primarily cultivated for export. Over the years, the tonnage has dramatically dropped. The primary pepper-producing state in India is Kerala. The chapter would discuss the various aspects of production in Kerala and pepper future markets, and there would also be details on the production of bush pepper.

Keywords India · Pepper · Economics · Pepper future market

Because India is the most important producer of black pepper in the world, a discussion on pepper economy and marketing patterns in the country is relevant. Currently, India accounts for 49.5% of the total acreage in the world, but in 1951, it was 70%. Though the country holds the largest pepper acreage in the world and contributes up to 29.8% of world production, its productivity is the lowest in the world—an abysmally low figure of 294 kg ha^{-1}. In fact, the total contribution of India to world production dropped from 66% in 1951 to the current 29.8%. This has been the case for exports as well. From 1951 to 1991, the India's share of the global pepper market fell from 56% to 23%. It is interesting to note that while India's position of preeminence in pepper production dwindled, other countries' performance was on the rise. The last decade of the past century witnessed a global downtrend in pepper production.

In India, pepper is primarily cultivated for export. From around 16,000 tons in 1950–1951, export has scaled up to 32,000 tons, a 100-fold increase, by the turn of the century. The 1990s witnessed a fluctuating market because of price fluctuations. The latter half of the 1990s witnessed a sharp increase in unit price, peaking at about US \$50 kg^{-1} in 1997, and though export per se witnessed a drop, it was more than compensated for by the escalating unit price. In the latter half of the 1990s, global pepper production was around 189,000 tons, of which more than 118,000 tons were exported as per the data provided by the IPC headquartered in Jakarta. In these estimates, India tops the list of producing countries with 65,000 tons, followed by

K. P. Nair, *The Geography of Black Pepper (*Piper nigrum*)*,
https://doi.org/10.1007/978-3-030-52865-2_11

Indonesia. India exports spices to more than 120 countries around the world; of these countries, export to 23 countries account for about 82% of the hard currency earned (in US $) by the country. In the case of pepper, India is the major supplier to 83 countries; among the importers, the USA tops the list with an α growth rate of 5% in quantity and 3% in value. Of the total spices imported into the USA, pepper accounts for about 15%, of which 50% originates in India. The USA is also the major importer of spice oils, primarily oleoresin, and India supplies close to 86%. However, the recent trend shows a decline in import into the USA. In the beginning of the last decade, pepper import from India to the USA was 56%, which declined to 30% by the middle of the decade.

Though the USA tops the list of pepper importers from India, it is interesting to note that per unit value realization is lowest from Russia, Canada, Italy, and Poland, with Russia topping the list. It is also interesting to note that there is a visible shift in the food habits of Americans, where the preference for hot spices like pepper in the last two decades accounts for nearly 75% of consumption. The US Trade Association projected spice consumption to peak to 1000 million pounds by the turn of the century (Cheriankunju 1996a). Maintenance of quality assumes the greatest importance in this regard. The United States Food and Drug Administration (USFDA) detained 1164 consignments of Indian spices due to contamination with seeds of noxious weeds in 1995. In fact, there has been a steep escalation of detection of contaminated consignments from 140 in 1991 to 757 by 1994 (Sivadasan 1996), which, indeed, is a very poor reflection of Indian trade. With the WTO tightening the noose on substandard-quality imported food articles from developing countries, India would do well to clean its own facilities, lest a very adverse fallout in the pepper market follows. It is, however, encouraging to note that the Spices Board of India, the apex body attached to the Ministry of Commerce, Government of India, has initiated a number of measures to ensure that quality of pepper is guaranteed both at the producer and the exporter levels.

The former USSR was the largest pepper importer from India, with a total tonnage of 19,400 in 1989–1990; the tonnage dropped dramatically to 6060 tons by 1994–1995 primarily because of political turmoil in the region (Cheriankunju 1996a). In spite of the economic instability, Russia continues to be a major consumer of pepper from India. Russia, with an import of about 3200 tons in 1994–1995, accounted for US $6.05 million, followed by Italy at US $3.62 million and Canada at US $2.54 million. Traditional export of pepper from India to such countries as Hungary, North Korea, Maldives, Trinidad, Yemen, the United Arab Emirates, and Yugoslavia has considerably decreased; in their place, new markets, such as Chile, Kazakhstan, Tajikistan, and Venezuela, have emerged. Of these, Kazakhstan imports the largest amount. Apart from these, Japan is fast emerging as a potential market for Indian pepper. This is mainly due to the fast-changing food habits of the country, where ready-to-eat fast foods are emerging along the lines of Western food habits, where pepper forms an important culinary component. In 1994–1995, Japan imported 41,911 metric tons of spices, of which 17.8% was pepper. Most of the pepper goes into industrial uses followed by domestic consumption and other service sectors. The annual rate of increase in spice import by Japan is estimated to be 5%

(Cheriankunju 1996b). The market share of Japan's import from India is stagnant at 5.5%, while that from Malaysia, the bulk exporter to Japan, stays at 68%. Indonesia contributes 31%. Indian pepper has a premium in the world market, and if strategic advantages are exploited following the establishment of the WTO, India can scale up its export to Japan. Additionally, Israel is a potential market for Indian pepper. In 1994–1995, Israel imported 6902 tons valued at US $7.42 million (Anon 1996). Among the potential markets for Indian pepper, the USA, Canada, Russia, the UK, and Japan are on the top rungs because these countries have a lower level of instability in pepper imports (Jeromi and Ramanathan 1983). According to the International Trade Center in Geneva, world trade in spices during the last decade averaged 450,000 metric tons valued at US $1500 million; pepper is the major component, which accounts for as much as 30–35% in quantity and 20–25% in value (Peter 1996).

11.1 Pepper Production Scenario in India

In 1950–1951, nearly 80,000 ha of cultivated area in India were under pepper, 98% in Kerala State. An analysis of the trends in growth with regard to area, production, and productivity is given in Table 11.1. It can be surmised from the table that during the last five decades, overall acreage, production, and productivity have declined all over India and, in particular, in Kerala State. The steepest fall was in productivity. On an all-India basis, while acreage slipped by only 22% from the 1950s to the 1990s, production and productivity declined by 54% and 89%, respectively. For Kerala, the respective figures for decline in area and productivity are 25% and 96%, while production increased by 70%. In fact, both on an all-India basis and for Kerala State, productivity showed very dramatic negative trends.

The reason for the steep fall in productivity, especially in Kerala, is due to the prevalence of diseases, of which the *Phytophthora* foot rot is the most serious; there is, as yet, no permanent remedy for this disease. As the question of area expansion has only very limited possibilities, if there is to be a breakthrough in pepper

Table 11.1 Area, production, and productivity growth rates for black pepper in India and Kerala State

Details		1950s	1960s	1970s	1980s	1990s	Overall
Area	India	0.87	1.12	1.05	1.81	1.11	0.68
	Kerala	0.94	0.19	−0.19	1.85	1.00	0.70
Production	India	1.36	−0.50	0.34	2.43	−0.32	0.63
	Kerala	0.40	−0.88	−0.40	2.26	0.70	0.68
Productivity	India	0.44	−1.40	1.27	0.63	−1.40	−0.65
	Kerala	0.46	−1.07	−0.2	0.41	−0.01	−0.02

Source: Economic Preview, Government of the State of Kerala
Note: Data given as log Y = a + bt in percent per annum

production in Kerala, it will have to be through breeding superior, disease-resistant varieties with good agronomy to match. Soil management is still based on book knowledge, and there is a very clear dearth of fresh ideas.

Pepper cultivation in Kerala is largely concentrated in certain districts. Kannur, Kozhikode, Kollam, and Thiruvananthapuram districts, in sequence, contribute more than 80% of the total pepper area in the state and 75% production of the state. By the early 1990s, the share of these four districts declined to 33% and 29% in acreage and production, respectively, and others took over. Currently, the districts of Idukki, Wayanad, and Kannur have emerged as the major pepper-producing districts of the state, contributing 60% of area and 67% of production. Among the three, Idukki District leads, with 21% area and 27% production. It is followed by the Wayanad District, which has contributed 25% of pepper produced in the state during the past decade. A very interesting development of late is the emergence of the Idukki District as the frontrunner in the production of "organic" pepper—much valued in the USA and Europe. Based on the cool climate of the region and relatively lower impact by the chemical agriculture, the district holds out great promise to emerge as a global leader in the production of organic pepper.

Pepper as a monocrop is confined to the districts of Idukki and Wayanad, which constitute only 3% of the total area (George et al. 1989). There is ample scope to cultivate pepper as an intercrop on coffee, tea, and cardamom plantations. The high-range districts of Kerala, such as Idukki and Wayanad, where currently most of these plantations are located, are potential areas where pepper can be grown as an intercrop. Silver oak, used as a shade tree in these plantations, can be used as standards (support for the pepper vine to trail). There is a strong case to initiate systematic field trials to develop a clear-cut package of practices in growing pepper as an intercrop on these plantations. Isolated studies have been carried out. However, no efforts on a large scale have been initiated. This is a very promising area for enhancing pepper production in India, but until now, it has not been tackled in a very systematic and scientific manner.

Pepper is an important component of "homestead farming" in Kerala, and the vines are trailed on coconut trees, areca nut trees, mango trees, jackfruit trees, and tamarind trees, which are generally grown on small farms owned by small-scale and marginal farmers. The level of inputs—namely, fertilizers, pesticides, and the like—in these situations is low. Transformation of these homestead gardens due to various socioeconomic reasons largely reduced this practice. Coupled with this, the incidence of foot rot swept away a substantial number of pepper plants on homestead farms. There is a strong case for developing a package of practices at low input levels in these homestead farms. There is a need to reintroduce pepper in homestead farming, both in urban and rural areas. For this purpose, bush pepper offers much promise.

11.2 Bush Pepper

Bush pepper holds out much promise in homestead farming. The plagiotropic, lateral fruiting branches of pepper exhibit sympodial growth, and these branches, when rooted and planted, grow into a bush. More fruiting branches are produced by these bushes, and flowering commences in the first year of planting itself. A common method developed at the Indian Institute of Spices Research (IISR) involves the following steps: (1) selecting healthy lateral shoots of the previous year's growth with two or three leaves, (2) dipping the cut end in a commercially available rooting hormone, and (3) planting them in moist rooting medium, which consists of weathered coir dust, in a 200-gauge PVC bag about 40 × 25 cm in size (later, the bags are kept under shade). Prior to closing the bag, air is blown into it. In about a month, the cuttings start producing roots. After about 2 months, the bags are opened and allowed to remain in that state for a few days before transplanting into clay pots of convenient size (Photo 11.1). Almost 80% success can be ensured provided the three steps are meticulously followed. Bush pepper can be planted at a minimum density of 2500 plants ha^{-1} spaced 2 × 2 m apart, and the population can be doubled if spacing is reduced. Record yields of 1960 kg ha^{-1} have been obtained using Panniyur-1 variety (Annual Report of IISR 1997), and yields in the range of

Photo 11.1 Production technology for *Trichoderma* (biological control of "quick wilt" caused by *Phytophthora* fungus)

Fig. 11.1 Bush pepper in India

1600–1700 kg ha^{-1} are quite common. Bush pepper is commonly grown in Indonesia, Malaysia, and the Philippines. It is still very far from catching on in India and the home state of pepper, Kerala. Figure 11.1 shows bush pepper growth in India.

11.3 Economics of Pepper Production in the State of Kerala

In India, Kerala is the most important pepper-producing state and merits further discussion. Vinod (1984) and Santosh (1985) have attempted to obtain a dependable estimate of the cost structure of pepper production in the Idukki and Kannur districts, respectively, which are at the forefront of pepper production in Kerala State. Pepper production in Kerala is highly labor-intensive, and labor costs account for more than 50% of the total cost of production. An important reason for the decline in pepper production in the state could be due to the nonavailability of timely labor and its very high wage structure on an all-India basis. Among all the states of India, Kerala's labor wages are the highest. Though Idukki District leads in cultivation, profitability is higher in Kannur District. The cost/benefit ratio was 1.09 for Idukki District and 1.16 for Kannur District as per the findings of Vinod (1984) and Santosh (1985). The payback period of pepper cultivation was estimated as 10 years in Idukki District and 11 years in Kannur District. The corresponding net present worth is approximately US $93 for Idukki District and US $148 for Kannur District, and the internal rate of return is 13.48% and 17.22%, respectively.

11.4 Marketing

Village traders dominate the local market inasmuch as small pepper farmers are concerned in Kerala. The trade is based, in many ways, on faith and long-term personal bonds, which is absent in the case of wholesale merchants who deal in large quantities. Additionally, transportation to the city or town market is quite often out of the reach of small farmers, and this also leads to the small farmer going to the village trader. By tradition, village traders collect the product directly from the farmers' yard or, if not possible, from neighboring village markets. Wholesale merchants, though, offer higher prices compared to the village trader. Village and internal wholesale merchants are important market intermediaries (Vinod 1984). Small and marginal farmers generally affect village sales in the nature of a preharvest contract, whereby the farmer is obliged to sell his product at a predetermined price prior to harvest; if there arises any subsequent fluctuation in price, both parties are bound to go by the earlier agreed-upon contract of sale.

The upcountry wholesale merchants transport the produce to Kochi (which is the port city of Kerala and the prime business center of the state), in the central part of the state. Once the produce reaches Kochi, the commission agents negotiate with the exporters or wholesale merchants or brokers. The brokers act on behalf of the internal wholesale merchants or exporters and negotiate with the commission agents. At the exporters' premises, the produce is processed and graded. Prior to the official clearance of the produce for export, it is subjected to check sampling and all grading tests. The consignment thus cleared for export is then handed over to the clearing or forwarding agency after completion of the paperwork. Subsequent work is carried out by the forwarding agency. Four major marketing channels were identified in Idukki District, while there were five in Kannur District. The share of the producer on the free on board (FOB) price of pepper was estimated to be 86.06%. Compared to this, the producer is the beneficiary of a higher share, which is 88.80%, when the produce is internally consumed and moved through the domestic consumption channel. In either case, it is clear that intermediaries corner a good percentage of the value chain.

11.5 Pepper Futures Market

The origin of the futures trade is almost a century old in India. Futures trade is an insurance against price fluctuations of commodities handled by traders. It was the Indian Pepper and Spice Trade Association (IPSTA) that heralded the futures trade in pepper in India in Kochi (Sethuram 1995). The move had a very positive effect on the transfer of risk and recovery of price. From hedgers, the risk is transferred to the speculators, who in turn provide the liquidity to the market.

Futures markets are standardized with regard to trading specifications and terms of delivery. The international pepper price shows a trend that is unfavorable to most of the pepper-producing countries. This is largely attributable to supply situations rather than production scenarios. The IPC, in an effort to ensure an efficient price management system, conducted a study on the viability of pepper futures contract, sponsored by the United Nations Conference on Trade and Development (UNCTA D). Based on this, a working group of representatives from IPC member countries was set up to examine the requirements of a global pepper futures contract. Consequently, the IPSTA at Kochi initiated action for the establishment of the International Pepper Exchange (IPE) at Kochi. The expectations are that it will provide an incentive in pepper trading, both in the domestic and international markets, ensure better prices to producers, and extend a hedging facility to exporters of all member countries. Further, it is also proposed to link the Kochi pepper exchange with the Kuala Lumpur Commodity Exchange at a future time, when a second trading floor would be started. The globalization process, in general, and the increasing volatility of pepper prices, in particular, necessitated the establishment of a global pepper market to trade pepper futures contract. This also facilitates a free trade regime in pepper, because the product can be freely moved among the countries without tariff barriers and quota restrictions.

Due to the additional premium for Indian pepper, its price in the international market is relatively higher than that of pepper produced by other countries. Low yield combined with high unit cost of production combined with better prices in the domestic market leads to a higher international price. Also, the improved stock-holding capacity of the farmers influences price, because, in times of price decline, farmers can hold on to their produce without unloading it to the market. Both corporate procurement and speculation play a significant role in pushing up the price. Futures trade ensures a constant supply and steady price for the produce.

One of the main criticisms against the exchange is how it translates the benefits to the actual producers. Pepper production in Kerala is mainly scattered among small farms, and the market awareness of these small and marginal farmers is indeed scanty. At least 72% of the farmers belong to this ignorant category (George et al. 1989). These small and individual farmers neither have sufficient quantity to trade nor the technical expertise to benefit from the pepper exchange. This has resulted in the futures trading in Kochi being handled mainly by representatives of exporters. The history of futures trading in pepper has clearly shown that producers have almost no say in its operations, including the large producers. The Primary Agricultural Credit Societies (PACS) have not attempted to enter the futures market either. The state marketing agencies, such as the State Trading Corporation and the State Marketing Federation, are wary of their association with the exchange setup because they have met with large financial losses. This is only natural—governmental agencies in Kerala rarely have true accountability, which is a legacy of the earlier socialistic pattern of governance in the state. If fair trading is to be ensured, the possibilities of manipulative tactics—which create an artificial market to the detriment of the interests of the producers—by the powerful members of the exchange have to be fully scotched. Invariably, however, the speculator calls the shots rather than the toiling producer, no matter how big or small. Despite these shortcomings, the IPSTA is hopeful of bridging the disparity between farm gate price and market price, often quite wide in certain instances, by eliminating needless intermediaries involved in transactions, which would ensure better prices for the producers. Liquidity of traders is expected to increase because banks are now willing to advance as much as 80–90% of the value of the produce when it is hedged against adverse price fluctuations through the medium of futures contract. Reckless speculations will be prevented by the imposition of limits on open position, daily price band fluctuations bound both up and down, and collection of special margin deposits to curb speculative activity through financial restraints.

An important component of the economics of pepper growing in Kerala is the availability of reliable market information to the farmers. As of now, most pepper farmers either have no dependable access to reliable market information or are illiterate and cannot make use of it. Hence, the need to educate farmers on the importance of acquiring reliable market information and marketing processes in pepper is of paramount importance. One of the prime reasons for pepper farmers' inability to realize a good price, especially in the case of small and marginal farmers, is the lack of adequate storage facilities. Holding capacity of the farmers can be enhanced through the provision of credit support. There is good scope to attempt a cooperative marketing system, as is the case of other plantation crops in the states (such as rubber, coconut, and areca nut). In the case of these crops, there are cooperative marketing societies spread across the state—a similar attempt for pepper could be initiated.

References

Annual Report (1997) Indian Institute of Spices Research, Vision 2020. IISR Perspective Plan, Calicut, Kerala State, India

Anon (1996) Israel—a promising market for Indian spices. Spice India 9:2–3

Cheriankunju NE (1996a) Know your market for spices—Japan. Spice India 9:5–7

Cheriankunju NE (1996b) Know your market for spices—USA. Spice India 9:2–5

George PS, Nair KN, Pushpangadan K (1989) The pepper economy of India. Oxford/IBH Publishing Co. Pvt. Ltd., New Delhi

Jeromi PD, Ramanathan A (1983) World pepper market and India: an analysis of growth and instability. Indian J Agric Econ 48:89–97

Peter KV (1996) Spices research and development-an updated overview. Empl News 21:1

Santosh P (1985) Cost of cultivation and marketing of pepper in Cannanore District. M.Sc. (Ag) thesis, Kerala Agricultural University, Thrissur, Kerala State, India

Sethuram P (1995) Cochin pepper market—a journey through history. Indian Spices 32:3

Sivadasan CR (1996) Import regulations in USA. Spice India 9:2–10

Vinod G (1984) Cost of cultivation and marketing of pepper in Idukki District. M.Sc. (Ag) thesis, Kerala Agricultural University, Thrissur, Kerala State, India

Chapter 12
Pepper Pharmacopoeia

Abstract The chapter discusses various aspects of pepper pharmacopoeia, with specific reference to its medicinal properties, It also discusses the role of pepper as an anti-inflammatory product and as a central nervous system depressant and its effects on hepatic enzymes, carcinogenic and mutagenic effect. Pepper is an antioxidant and an antimicrobial agent and has toxicological agent affects. The overall pharmacological effects of pepper in human health are aspects discussed in detail in the chapter.

Keywords Pepper pharmacopoeia · Inflammation · Central nervous system · Antimicrobial · Antioxidant

The use of black pepper as a drug in both the Indian and Chinese systems of medicine, apart from its wide-spectrum use as a food additive, is well documented (Atal et al. 1975; Kurup et al. 1979; Nadkarni 1976). Pepper is elaborately described for its medicinal values in the ancient Indian system of medicine *Ayurveda*. It is described as *katu* (pungent), *tikta* (bitter), and *ushnaveerya* (potent). Further mention of pepper is also made with regard to the control of the principal sources of diseases (as described in *Ayurveda*—namely, *vata* and *kapha*, in the ancient Indian language Sanskrit) as the biological processes in the human system controlled by the CNS and the later formation and regulation of all fluids that have a preservative quality (such as mucus, synovia, and the like). Pepper is described as a drug that enhances digestive power in the body, improves appetite, and cures colds, coughs, dyspnea, diseases of the throat, intermittent fever, colic, dysentery, worm infestation, and hemorrhoids. Pepper is also prescribed to relieve toothache, muscular pain, inflammation, leucoderma, and even epileptic fits (Ayier and Kolammal 1966; Kirtikar and Basu 1975). Black pepper is called *maricha* or *marica* in Sanskrit, which implies its ability to dispel poison. In the Chinese system of medicine, pepper is used as an antidote against snake and scorpion bites.

The preceding descriptions explain the diverse actions of pepper being used in the Indian system of medicine, either as such or through its active ingredients in many

K. P. Nair, *The Geography of Black Pepper (*Piper nigrum*)*,
https://doi.org/10.1007/978-3-030-52865-2_12

formulations. In sharp contrast to the use of pepper in the Indian system of medicine, pepper finds no use in allopathy or homeopathy. The beneficial effect of pepper in medical formulations can be attributed to piperine and other phenolic amides and essential oils. But in *Ayurveda*, the active ingredient-based specific activities are not taken into consideration. Hence, the pharmacological and toxicological aspects of pepper and its constituent secondary metabolites were not studied. It must be, without doubt, mentioned that many of the clinical uses of pepper are time-tested, over several generations, and there is implicit faith among the Indian masses on the unquestionable medicinal value of pepper. However, to scientifically establish the value of pepper as a unique medicinal product, it is crucial to investigate the pharmacological aspects of pepper and its active ingredients based on well-defined experimental protocols currently used in modern drug research and development. Some of the preliminary investigations carried out over the past two decades in India on these lines provide valuable information. Antipyretic, analgesic, anti-inflammatory, antimicrobial, and antineoplastic activities were reported after in vitro and in vivo investigations. Of late, there is a surge of interest in environmentally friendly use of products to control crop pests, and pepper has been found to possess insecticidal as well as insect-repellent properties. Developments of new pepper-based products as insecticides could be a boon to human use because they would be free from the toxic nature of most of the insecticides currently in use. Piperine is the major alkaloidal constituent of pepper. Systematic pharmacological studies on piperine have revealed its analgesic (pain relieving), antipyretic (fever relieving), and anti-inflammatory properties, in addition to rejuvenation of the CNS. The antimicrobial activities are provided mainly by the essential oils.

In addition to the ancient *Ayurveda* system of medicine, there are other native systems of medicines in India, such as *Siddha* and *Yunani*. Pepper is used in these as well. In the Chinese system of medicine, pepper and chenghan (*Dichroa febrifuga* L.) are used in the treatment of malaria (Das et al. 1992). The active ingredients of chenghan are febrifugine and isofebrifugine, both of which show a 50% higher antimalarial property than the conventional quinine. As an antipyretic, pepper is also used in the treatment of malaria (Nadkarni 1976). The curative effects claimed in the preceding cases with regard to pepper can be attributed to the antipyretic and analgesic actions of the active ingredients. A strong antipyretic effect on typhoid vaccinated rabbits at a dose of 30 mg kg^{-1} body weight after oral administration has been reported by Lee et al. (1984). In these studies, acetaminophen was used as the reference drug. The antipyretic action of piperine was found to be stronger than that of acetaminophen. To determine the analgesic action, acetaminophen and aminopyrine were used as reference compounds. Piperine showed a strong activity, with an LD_{50} value of 3.7 mg kg^{-1} dose using the writhing method and 104.7 mg kg^{-1} using the tail-clip method.

12.1 Anti-inflammatory and Central Nervous System Depressant Activity of Pepper

Several researchers have reported the anti-inflammatory property of pepper. Piperine showed a significant inhibition of the increase in edema volume in a carrageenan-induced test with an oral intake of 50 mg kg^{-1} body weight (Lee et al. 1984). The effect of different acute and chronic experimental models was studied by Mujumdar et al. (1990). These researchers evaluated the mechanism of anti-inflammatory activity by biochemical processes and concluded that piperine acted positively in the case of early acute changes in inflammatory processes. These studies have been further corroborated by those of Kapoor et al. (1993).

In the case of epileptic patients, pepper is used in *Ayurveda* to induce sleep (Kirtikar and Basu 1975). In pharmacological investigations, pepper and piperine were found to have a CNS depressant effect. Both pentetrazole seizure and maximal electroshock seizure were inhibited by pepper and piperine (Won et al. 1979). The LD_{50} value of 287.1 mg $kg-^{1}$ i.p. was found for piperine and a corresponding oral value was 1638.8 mg $kg-^{1}$. In another investigation, Shin et al. (1980) showed that piperine at 1/10, its LD_{50} value had a strong potentiating effect on hexobarbitol-induced hypnosis in mice. Decreased passivity, ptotic symptoms, and decrease in body temperature were also observed. Protection against electroshock seizure and a muscle relaxant effect were observed at a relatively low intraperitoneal dosage range of an LD_{50} of 15.1 mg $kg-^{1}$ (Lee et al. 1984). This dosage appeared almost equipotent to the reference compound phenytoin. Petrol extract of pepper leaves was found to potentiate phenobarbitone-induced hypnosis in mice (Sridharan et al. 1978). Majumdar et al. (1990) also reported that a high dosage of piperine potentiates the phenobarbitone sleeping time by inhibiting the liver microsomal enzymes. It also acted partially through stimulation of the pituitary adrenal axis.

The preceding investigations clearly show that piperine has the capacity to be an effective pepper ingredient in the treatment of petit mal (minor ailments). Antiepilepsirine (AE) isolated from white pepper is a compound of very great pharmacological interest. AE is 1-(3-benzodioxol-5-yl)-1 (oxo-2-propenyl)-piperidide. Perhaps this is the only compound derived from pepper that is clinically used in the treatment of human epilepsy. This is an alternative to Dilantin therapy, which is used in Chinese hospitals for the treatment of epilepsy (Ebenhoech and Spadaro 1992). It is probable that AE may release 5-hydroxytryptamine (5-HT) from nerve endings, which can intensify the anticonvulsive state (Liu et al. 1984). Studies using rats as test animals show that AE increases 5-HT concentration in their brains, intensifying the anticonvulsive state. AE also raises the tryptophan level in the brain, which causes elevation of serotonin and monoamine levels, and leads to control of seizures.

12.2 Effect on Hepatic Enzymes

While pepper or piperine positively affects liver function, as evidenced by the research of many, they in no way cause hepatic toxicity. Piperine functions mainly as a chemopreventive substance through modulating enzyme function. Hepatic toxicity of piperine in rats was studied by estimating the hepatic mixed function oxidases and serum enzymes as specific markers of hepatotoxicity (Dalvi and Dalvi 1991). An intragastric dose of 100 mg $kg-^{1}$ body weight resulted in an increase in hepatic microsomonal enzymes 24 h after treatment. In the experiment, cytochrome P450, cytochrome b5, NADPH-cytochrome C reductase, benzphetamine N-demethylase, aminopyrine N-demethylase, and aniline hydroxylase were estimated. An intraperitoneal dose of 100 mg $kg-^{1}$ body weight did not produce any effect on the activity of the drug-metabolizing enzymes. However, when the dose was enhanced an eightfold to 800 mg $kg-^{1}$ body weight as intragastric and 100 mg $kg-^{1}$ body weight as intraperitoneal, a significant decrease in the levels of the enzymes was noted. These treatments did not, however, affect those serum enzymes that are specific markers of toxic liver conditions. Piperine has been found to provide significant protection against chemically induced hepatotoxicity. Both in vitro and in vivo lipid peroxidation and prevention of depletion of GSH and total thiols were observed (Kaul and Kapil 1993). Liquid peroxidation causes the production of free radicals, which in turn causes tissue damage. GSH conjugates xenobiotics, which are excreted by subsequent glucuronidation. In the previously cited study, the hepato-protective action of piperine was compared with a reference compound, silymarin, a known hepato-protective drug, and it was found that piperine has slightly lower activity. In a feeding experiment using Swiss albino mice fed with a diet containing 1, 2, and 3% black pepper, w/w, for 10 and 20 days, a dose-dependent increase in the levels of the hepatic biotransformation enzymes—namely, glutathione-s-transferase, cytochrome P450, cytochrome b5, and acid-soluble sulfhydryl-s—was observed (Singh and Rao 1993). Reen et al. (1993) found lowered levels of glucuronidation due to the inhibition of the enzyme UDP-glucose dehydrogenase in an in vitro study. Tripathi et al. (1979), while investigating the hypoglycemic action of several plants, found that pepper fruits are devoid of any significant hypoglycemic action in rabbits. The aqueous extract of pepper leaves at a dose of 10–20 mg $kg-^{1}$ body weight led to a moderate increase in the blood pressure of dogs (Sridharan et al. 1978). Both piperine and AE have detoxifying qualities that may lead to an increase in the bioavailability of other drugs, as a result of which they alter the pharmacokinetic parameter of the epileptic (Bano et al. 1969).

12.3 Carcinogenic and Mutagenic Effects of Black Pepper

Positive effects of pepper as an anticarcinogenic and nonmutagenic source have been found by investigators. The Ames test shows its nonmutagenic quality. Chemical carcinogenesis has been found to be prevented by pepper because of its stimulatory effect on xenobiotic biotransformation enzymes. The antioxidant properties of piperine and associated unsaturated amides play a preventive role in carcinogenesis. Dietary intake of natural antioxidants can be a crucial aspect of building up the human body's natural defense mechanisms against the onslaught of diseases, which can be caused by degradative changes brought about by mutagens. Additionally, the essential oil constituents in pepper inhibit DNA adduct formation by xenobiotics. This observation shows that pepper has anticarcinogenic potential. However, pepper extracts showed enhanced incidence of tumor formation in mice and an elevated level of DNA damage caused by piperine in cell culture investigations. Hexane, water, and alcohol extracts of pepper were tested for mutagenicity on *Salmonella typhimurium* strains TA 98 and TA 100 by Ames assay, and the results indicated that the extracts were nonmutagenic. This investigation provides evidence that water extract exerts an antimutagenic action on carcinogen-induced mutagenesis (Higashimoto et al. 1993). The chemoprotective role of pepper, as well as its constituents, has been indicated because of its positive effect on the activity of biotransformation enzymes in the liver in a dose-dependent manner (Singh and Rao 1993).

Nakatani et al. (1986) report a very significant antioxidant activity due to the phenolic amides present in pepper. It is clearly established that antioxidants exert a preventive role in carcinogenesis. The modulating effects of the essential oil constituents of pepper were investigated by Hashim et al. (1994), who showed that they have an inhibitory effect on carcinogenesis. The volatile oil and its constituents suppress the formation of DNA adducts with aflatoxin B1. The microsomal enzymes modulate this action. Feeding powdered pepper to mice at 1.66 w/w dose showed no positive impact on carcinogenesis, while feeding and painting of the mice body with solvent extract of pepper, 2 mg for 3 days per week for 3 months, showed an increased incidence of tumor formation (Shwaireb et al. 1990). The activity of methylcholanthrene, a potent carcinogenic compound, was found to be reduced by a pepper terpenoid d-limonene (Wuba et al. 1992). Two minor constituents of pepper—safrole and tannic acid—have minor carcinogenic activity.

In a tissue culture study using V-79 lung fibroblast cell lines, it was found that piperine-treated cell lines showed increased DNA damage compared with untreated ones (Chu et al. 1994). Piperine treatment lowered the activity of the enzymes glutathione-s-transferase and uridine diphosphate glucuronyl transferase, indicating cytotoxic potential. The in vivo formation of N-nitroso compounds from naturally occurring amines and amides contribute to the carcinogenic potential of certain foods and food additives. Piperine and other phenolic amides present in pepper are also known for their conversion to N-nitroso compounds in acidic conditions and, hence, are treated as carcinogenic (Lin 1986). However, it can be inferred that the presence of a conjugated unsaturated system in the phenolic amide prevents the oxidation of the amide nitrogen to N-nitroso compounds to a great extent. Additionally, the essential oil constituents in pepper would also ensure DNA stability because of their anticarcinogenic potential.

All of the cited research results point to the ambivalent nature of pepper—both as an anticarcinogenic agent and as a procarcinogenic agent. But overwhelming evidence indicates that it is more of the former than the latter. It is because of this belief that pepper has been a crucial ingredient of the Indian systems of medicine, primarily *Ayurveda* and to a lesser extent *Siddha* and *Yunani*, for centuries, and continues to be so even today.

12.4 Pepper as an Antioxidant

Antioxidants are one of the most crucial biochemical compounds in the human system ensuring good health. They scavenge free radicals, which trigger many untoward biochemical reactions in the human system, and control lipid peroxidation in mammals. Of late, there is a considerable interest in antioxidants. Lipid peroxidation is a chain reaction that is a constant source of free radicals, which initiate further peroxidation and trigger deterioration of food but also damage tissue (both of animal and plant origin), which in human beings causes many inflammatory diseases, aging, atherosclerosis, and cancer. Many investigations reveal that pepper and its phenolic constituents, such as amides, possess good antioxidant properties. Tocopherol and vitamin C are two important natural antioxidants.

Chiapault et al. (1955) have investigated the antioxidant property of spices in a two-phase aqueous fat system. The investigations indicated that pepper has an antioxidant index of 6.1, while turmeric and clove show values of 29.6 and 103.0, respectively. The antioxidant property of pepper has been attributed to its tocopherol content (Saito and Asari 1976). Revankar and Sen (1974), however, attribute the antioxidant property of pepper to its polyphenolic content. These authors investigated the effect of pepper oleoresin on fish oil and arrived at this conclusion. Abdel-Fattah and EI-Zeany (1979) observed that the autoxidation of unsaturated fatty acids and proteins was delayed by the addition of pepper and substantial protection against oxidative degradation was obtained. Nakatani et al. (1986), while investigating the family Piperaceae, established the fact that the five phenolic amides present in *P. nigrum* possess very good antioxidant properties, which were found to be superior to synthetic antioxidants, such as butylated hydroxytoluene and butylated hydroxyanisole.

12.5 Pepper as an Antimicrobial Agent

Volatile oils, which are an active constituent of most spices (and of which pepper is the most important), impart antiseptic, antibacterial, and antifungal properties. The positive effect of volatile oils can be attributed to their terpenoid constituents. Pepper possesses both bactericidal and bacteriostatic properties. These properties aid in enhancing the shelf life of foods to which pepper has been added. Even the leaf extract of pepper possesses the antibacterial activity (Subramanyam et al. 1957). However, the extract was not inhibitory to the growth of *E. coli*, *Aerobacter aerogenes*, *Lactobacillus casei*, *Staphylococcus faecalis*, *S. aureus*, and *S. sonnei* (Subramanyam et al. 1957). The essential oil obtained from pepper was found to be inhibitory to the growth of a penicillin-C-resistant strain of *S. aureus*. Jain and Kar (1971) documented the inhibitory action of pepper oil on *Vibrio cholerae*, *S. albus*, *Clostridium diphtheriae*, *Shigella dysenteriae*, *Streptomyces faecalis*, *S. pyogenes*, *B. pumilus*, *B. subtilis*, *Micrococcus* sp., *Pseudomonas pyogenes*, *P. solanacearum*, and S. typhimurium. The antibacterial action was determined by the agar well-diffusion technique using cefazolin as standard (Perez and Anesini 1994). The mycelial growth and aflatoxin synthesis by *Aspergillus parasiticus* were inhibited by pepper oil at a concentration of 0.2–1% (Tantaoui and Beraoud 1994). Antifungal activity against *Candida albicans* (Jain and Jain 1972) was exhibited by pepper leaf oil. Rao and Nigam (1976) reported a similar effect of pepper leaf oil on *A. flavus*. The antibacterial effects of pepper extract, essential oil, and isolated piperine in vitro against sausage microflora, *L. plantarum*, *Micrococcus* specialis, and *Streptococcus faecalis* have been reported by Salzer et al. (1977), in which the authors noted that only isolated piperine displayed microbial growth inhibiting effects at a normal dose. Pepper powder and extract were active only at high concentrations. An alcoholic extract of pepper was found active against the deadly foodborne bacterium *C. botulinum* (Huhtanen 1988). Another potential activity was revealed by a study carried out using tincture of pepper by Houghton et al. (1994). These authors investigated the antibacterial activity against nine strains of *Mycobacterium tuberculosis* and found that growth of all nine strains was inhibited.

The mode in which pepper acts positively is through its effect on enhancing the bioavailability of administered medicaments, when they contain pepper or its active ingredients, and uptake of proteins and amino acids from ingested food. In an investigation using *Trikatu*, which is an *Ayurvedic* preparation containing *P. nigrum*, *P. longum*, and turmeric (*Zingiber officinale*, another spice with great therapeutic value), it was observed that the pepper-containing preparation enhances the bioavailability of other medicaments (Johri and Zutshi 1992). At a dose of 25–250 μ mol$-^{1}$, piperine enhanced the uptake of l-leucine, l-isoleucine, and l-valine and increased lipid peroxidation in an in vitro study using intestinal epithelial cells of rats (Johri et al. 1992). Piperine probably interacts internally to enhance the permeability of intestinal cells. Protein uptake from pulses was enhanced by up to 1.5% with the addition of pepper (Pradeep and Geervani 1994). The study mainly involved two pulses in the Indian vegetarian diet—winged bean and horse gram. The results clearly suggest that the Indian vegetarian diet, where spice is a common ingredient, has a decided advantage in human health.

12.6 The Pharmacological Effect of Pepper on Human Health

The precise role of pepper on human health can only be understood through well-structured studies, which are only very limited in number to date. However, from the preceding sections, it can be concluded that where extrapolations and comparisons can be made, pepper (as a whole product) or its active ingredients have been found to have a very positive role on human health. A brief description of the limited number of studies is given here.

When used in high doses, gastric mucosal injury caused by pepper is comparable with that of aspirin. This observation was made in a double-blind study of intragastric administration of pepper to human volunteers (Mayore et al. 1987). In this investigation, healthy human volunteers were given meals containing 1.5 g of pepper; 655 g of aspirin and distilled water were used as positive and negative controls. This was a short-term study, and it must be pointed out that long-term effects of daily pepper intake, through food or medicaments, are unknown. Vezyuez Olivincia et al. (1992) investigated the effect of intestinal peristalsis by measuring the orocecal transit time (OCTT) utilizing a lactulose hydrogen breath test on healthy human subjects. They were given 1.5 g of pepper in gelatine capsules, and the OCTT was measured after several days; it was found that OCTT increased significantly after administration of pepper. This finding has great clinical importance in the management of various gastrointestinal tract disorders. An equally important investigation reveals the effectiveness of pepper extract volatiles in the treatment of cessation of smoking. Results of the investigation on human subjects reveal that cigarette substitutes, which deliver pepper volatile compounds, alleviated smoking withdrawal symptoms. The results of both the investigations detailed have very great potential for further experimentation and, perhaps, pharmacological exploitation for further therapeutic benefits for human health.

12.7 Clinical Applications of Pepper

It is only in the case of the Indian system of medicine—primarily *Ayurveda* and, to a lesser extent, *Siddha* and *Yunani*—that pepper has been used in many clinical applications. No such use is made of pepper in allopathy. When clinically used, none of the active pepper ingredients are used as such. Dried black pepper with other medicinal plants, also in the dried form, is used in the preparation of specific formulations. The most widely used formulation in *Ayurveda* is

Dasamulakatutrayadi Kashayam, a formulation that is used in the treatment of asthma, and then comes *Ashtacurnam*, which is used to treat dyspepsia, flatulence, and the like. The formulation also has stomachic and carminative effects. *Amritharishtam* is used as an antipyretic and also in the case of women with excessive menstrual bleeding. *Muricadi Thailam* is used as an anti-inflammatory agent and also to relieve rheumatic pain. To cure coughs and bronchitis, *Dasamoola Rasayanam* is used, and *Guggulutiktaka Ghritham* has both analgesic and anti-inflammatory properties. Though pepper as such is widely used in the preparation of *Ayurvedic* medicines, its antipyretic, analgesic, and anti-inflammatory properties merit further research for use in allopathy, because, to date, pepper is not used in the preparation of any of the allopathic medicines. Its antioxidant and antimicrobial activities are also worth investigating further for possible use in the manufacture of allopathic medicines. Additionally, pepper has a good dietary value. It has a high fiber content (15–33%), high iron content (54–62 mg g^{-1}), high calcium content (1–1.5%), and also appreciable amounts of essential amino acids (Uma Pradeep et al. 1993). As a good digestive, pepper enhances the secretion and flow of the salivary enzyme amylase.

12.8 Toxicological Effects

Published reports do not detail any toxic effects of pepper. This may be due to the relatively small amounts used in many medicinal formulations. Because, in most cases, the total contents of piperine and associated phenolic amides used would add up to only 7–9%, w/w, and that of the volatile oils 2–4%, which are negligible amounts, no untoward health hazard has been encountered so far. At this level, the actual doses of the different constituents available from the quantity of pepper powder, oleoresin, or extractive used will be very unlikely to trigger any toxic side effects in human body. In fact, the very pungency of piperine and the strong flavor of the volatile oils act as deterrents against excessive human consumption.

The FAO (Food and Agriculture Organization) and WHO Expert Committee on Food Additives do not prescribe any limit to acceptable intake of piperine and the volatile oils. The major untoward consequence of pepper use is gastric mucosal injury at a dose of 1.5 g $kg-^{1}$ of food. It enhances the DNA adduct formation. The extract of pepper produces enhanced cancer incidence in mice. When mice were fed with extract of black pepper, it resulted in the formation of hepatocellular carcinoma, lymphosarcoma, and fibrosarcoma (El-Mofty et al. 1991). However, a large number of investigations show that pepper has anticarcinogenic attributes.

Before conclusive evidence can be arrived at—whether pepper and its active ingredients trigger anticarcinogenic or procarcinogenic reactions in human beings or not—much more detailed scientific scrutiny needs to be carried out. As of now, such studies are conspicuously lacking. The observation that pepper enhances the liver microsomal enzyme activity must also be investigated further. This will help decide whether or not constant use of pepper in the human diet leads to production of nonspecific enzyme induction. Pepper is a universal dietary component on the Indian subcontinent, and the quantity used is much more than used anywhere else in the world. Such information has very vital implications of health for a population that currently stands at more than a billion.

12.9 The Insecticidal Activity of Pepper

The primary advantage of using plant-derived insecticides in agriculture is that while the product is toxic to the pest in question, it is nontoxic to human beings (including those who handle it in the field), unlike most other chemically produced commercial insecticides, which have come under attack from the environmentalists. Plant-derived products have been used since time immemorial both in agriculture and for domestic purposes. The extract of pepper, volatile oil components, and the different phenolic amides present in pepper have shown insecticidal, insect repelling, and ovicidal activities to various plant insects and pests harmful to human beings. The major alkaloid of pepper—piperine—is more toxic than pyrethrine, a standard insecticide for house flies (Harvill et al. 1943). In a number of instances, the insecticidal or insect-repellent activity was obtained at low concentrations. Hence, this aspect needs to be further probed for positive field application.

The volatile oil of pepper was found to cause mortality of the cigarette beetle (*Lasioderma serricorne*) (Samuel et al. 1984). Several groups of researchers investigated the effectiveness of the alcohol extract of pepper and found it effective against cotton boll weevil, rice weevil, and *Drosophila* (Barakat et al. 1985; Scott and McKibben 1978; Su 1977). The LD_{50} value for the tropical application of the crude and purified extracts, obtained after 24 h, showed a mortality rate of 3.4 and 4.8 μg, respectively, per adult insect of *Sitophilus oryzae* and 4.5 and 7.2 μg, respectively, per adult insect of *Callosobruchus maculatus*. Oleoresin, piperine, and other amides in pepper were found to be lethal to the culex mosquito (*Culex pipiens*) and also to the pulse weevil (Su 1977). Pepper amides, such as piperonal, piperine, piperoline, pellitorine, pipercide, and dihydropipercide, also were investigated by Miyakado et al. (1979), Masakazu et al. (1980), and Deshmukh et al. (1982) for their effectiveness against insects and were found to be effective against the pulse beetle, rice weevil, and *Drosophila*, where larval growth inhibition was found. Desai et al. (1997) showed that 100% mortality within 24 h at a dose of 80 ppm was obtained using acetone extract of pepper in the case of *Anopheles subpictus* larvae.

References

Abdel-Fattah IE, El-Zeany BA (1979) Effect of spices on the autooxidation of fatty foods. Rev Ital Soztanze Grasse 56:441–443

Atal CK, Dhar KL, Singh J (1975) The chemistry of Indian Piper. Lloydia 38:250–264

Ayier KN, Kolammal M (1966) Pharmacognosy of ayurvedic plants, vol 9. Kerala University, Trivandrum, p 63

Bano G, Alma V, Raina U, Zutshi V (1969) The effect of piperine on the pharmokinetics of phenytoin in healthy volunteers. Planta Med 12:566–569

Barakat AA, Family HSM, Kandil MA, Ebrahim NMM (1985) Toxicity of the extract of black pepper, cumin, fennel, chamomile and Lupin against *Drosophila ceratetis* and *Spodoptera*. Indian J Agric Sci 55:116–120

Chiapault JR, Mizuna GR, Hawkins JM, Lundberg WO (1955) The antioxidant proper-ties of spices in oil-in-water emulsions. Food Res 20:443–451

Chu CV, Chang JP, Wang CJ (1994) Modulatory effect of piperine on benzopyrene cyto- toxicity and DNA adduct formation to V-79 lung fibroblast cells. Food Chem Toxicol 39:373–377

Dalvi PR, Dalvi PS (1991) Comparison of the effect of piperine administered intragastri- cally and intraperitonially on the liver and liver mixed function oxidases in rats. Drug Metabol Drug Interact 9:23–30

Das PC, Das A, Mandal S, Chatterjee A (1992) On the validity of the ethnic use of *P. nigrum* and *M. indica* L. In: Proceedings of the international seminar on traditional medicine, Calcutta, India, p 84

Desai AE, Ladhe RU, Deorag BM (1997) Larvicidal activity of acetone extract of *Piper nigrum* seeds to a major vector, *Anopheles subpietus* from Nashik region. Geobios 24:13–16

Deshmukh PB, Chavan SR, Renapurkar DM (1982) A study of insecticidal activity of twenty indigenous plants. Pesticides 16:7

Ebenhoech A, Spadaro O (1992) Antiepilepsirine: a new Chinese anticonvulsant herb drug. J Eco Tax Bot 16:99–102

El-Mofty MM, Khudoley VV, Shwaireb MH (1991) Carcinogenic effect of force-feeding an extract of black pepper in Egyptian toads. Oncology 48:347–350

Harvill EK, Hartzell A, Arthur JM (1943) Toxicity of piperine solution to house flies control. Boyce Thompson Inst 13:87–92. Merck Index, 1976, p 1266

Hashim S, Aboobaker VS, Madhubala R, Bhattacharya RK, Rao AR (1994) Modulatory effects of essential oils from spices on the formation of DNA adducts by aflatoxin B *in vitro*. Nutr Cancer 21:169–175

Higashimoto M, Purinatrapiban J, Katoka K, Kinouchi T, Vinitkekumnuen U, Akimoto M, Dinishi M (1993) Mutagenicity and antimutagenicity of extracts of three medicinal plants. Mutat Res 303:135–142

Houghton PJ, Astaniou A, Grange JM, Yates M (1994) Antibacterial effects of extracts and constituents of *Piper nigrum* L. Pharm Pharmacol 46(Suppl 2):1042

Huhtanen CN (1988) Inhibition of *Clostridium botulinum* by spice extracts and aliphatic alcohols. J Food Prot 43:195–196

Jain SR, Jain MR (1972) Antifungal studies on some indigenous volatile oils and their combinations. Planta Med 22:136

Jain SR, Kar A (1971) Antibacterial activity of some essential oils and their combinations. Planta Med 22:118

Johri RK, Zutshi U (1992) An Ayurvedic formulation of "Trikatu" and its constituents. J Ethnopharmacol 37:85–91

Johri RK, Thusu N, Khajuria A, Zutshi U (1992) Piperine mediated changes in perme- ability of rat intestinal epithelial cells. Biochem Pharmacol 43:1401–1407

Kapoor VK, Chawla AS, Manoj K, Pradeep K (1993) Search for antiinflammatory agents. India Drugs 30:481–493

Kaul IB, Kapil A (1993) Evaluation of liver protective potential of piperine: an active prin- ciple of black pepper. Planta Med 59:413–417

Kirtikar KR, Basu BD (1975) Indian medicinal plants III. Bishan Singh Mahendrapal Singh, Dehra Dun, pp 2133–2134

Kurup PNV, Ramdas VNK, Joshi P (1979) Handbook of medicinal plants. Indian Council of Agricultural research, New Delhi, p 143

Lee EB, Shin KH, Woo WS (1984) Pharmacological study of Piperine. Arch Arch Pharm Res 7:127–132

Lin JK (1986) Food-borne amines and amides as potential precursors of endogenous car- cinogens. Proc Natl Sci Council Rep China (B) 10:20–34

Liu G et al (1984) Quoted from Ebenhoech and Spadaro (1992)

Majumdar AM, Dhuley JN, Deshmukh VK, Raman PH, Thorat SL, Naik SR (1990) Effect of piperine on pentobarbitone induced hypnosis in rats. Indian J Exp Biol 28:486–487

Masakazu M, Isamu N, Hirosuki Y (1980) The Piperaceae amides part III. Insecticidal joint action of pipercide and co-occurring compounds isolated from *Piper nigrum*. Agric Biol Chem 44:1071–1073

Mayore M, Smith JL, Garhaur DY (1987) Effect of red pepper and black pepper on the stomach. Am J Gastroenterol 82:211–214

Miyakado M, Nakayama I, Yoshoka H, Nakatani NN (1979) The Piperaceae amides: structure of pipercide. Agric Biol Chem 43:1609–1611

Mujumdar AM, Dhuley JN, Deshmukh VK, Raman PH, Naik SR (1990) Antiinflammatory activity of piperine. Jpn J Med Sci Biol 43:95–100

Nadkarni KM (1976) Indian Materia Medica. Popular Prakashan, Bombay, pp 971–972

Nakatani N, Inatani R, Ohta H, Nishioka A (1986) Chemical constituents of pepper and application to food preservation naturally occurring antioxidant. Environ Health Perspect 67:135–147

Perez C, Anesini (1994) Antibacterial activity of elementary plants against *Staphylococcus aureus* growth. Am J Chin Med 22:169–174

Pradeep KU, Geervani P (1994) Influence of spices on protein utilisation of winged bean and horsegram. Plant Foods Hum Nutr 46:187–193

Rao CSS, Nigam SS (1976) Antimicrobial activity of some Indian essential oils. Indian Drugs 14:62

Reen RK, Jamwal DS, Taneja SC, Koul JL, Dubey PK, Wiebel FJ, Singh J (1993) Impairment of UDP-GDH and glucuronidation activities in liver and small intestine of rat and Guinea pig *in vitro* by piperine. Biochem Pharm 46:229–238

Revankar GD, Sen DP (1974) Antioxidant effects of a spice mixture on sardine oil. J Food Sci Technol (India) 11:31–32

Saito Y, Asari T (1976) Studies on the antioxidant properties of spices. I. Total tocopherol content in spices. J Jpn Soc Food Nutr 29:289–292

Salzer UJ, Broker U, Klie HP, Liepe HU (1977) Effect of pepper and pepper constitu- ents on the microflora of sausage products. Fleischwirtschafe 57:2011–2121

Samuel R, Prabhu VKK, Narayanan CS (1984) Influence of spice essential oils on the life history of *Lasioderma serricorne*. Entomon 9:209

Scott WP, McKibben GH (1978) Toxicity of black pepper extract to ball weevil. J Econ Ent 71:343–344

Shin KH, Yun HS, Won WS, Lee CK (1980) Pharmacological activity of *Piper retro- fractrum*. Soul Tacha Sean Young Opin 18:87–89. (Chem. Abstr. 93, 215484s)

Shwaireb MH, Waba H, El-Mofty MM, Dutter A (1990) Carcinogenesis induced by black pepper and modulated by vitamin-A. Exp Pathol 40:233–238

Singh A, Rao AR (1993) Evaluation of the modulatory influence of black pepper on the hepatic detoxification system. Cancer Lett 72:5–9

Sridharan K, Kalla AK, Singh J (1978) Chemical and pharmacological screening of *Piper nigrum* L. leaves. J Res Indian Med Yoga Homeopath 13:107

Su HCF (1977) Insecticidal properties of black pepper on cotton weevil and cow-pea weevil. J Econ Ent 70:18–21

Subramanyam V, Sreenivasamoorthy N, Krishnamoorthy K, Swaminathan M (1957) Studies on the antibacterial effects of spices. J Sci Ind Res 16:240

Tantaoui EA, Beraoud L (1994) Inhibition of growth and aflatoxin production in *Aspergillus parasiticus* by essential oils. J Environ Toxicol Pathol Oncol 13:67–72

Tripathi SN, Tiwari CN, Upadhyaya BN, Singh KRS (1979) Screening of hypoglycae- mic action in certain indigenous drugs. J Res Ind Med Yoga Homeopath 14:159

Uma Pradeep K, Geervani P, Eggum BO (1993) Common Indian spices: nutrient composi- tion, consumption and contribution to dietary values. Plant Foods Hum Nutr 44:137–148

Vezyuez Olivincia W, Shah P, Pitchumoni CS (1992) The effect of red and black pepper on orocecal transit time. J Am Coll Nutr 11:228–231

Won WS, Lee EB, Shin KH (1979) CNS depressant activity of piperine. 18:66–70. (Chemical Abstracts. 93, 215482q)

Wuba H, El-Mofty MM, Schweureb MH, Dutter A (1992) Carcinogenicity testing of some constituents of black pepper. Exp Pathol 44:61–65

Chapter 13
Consumer Products Out of Black Pepper

Abstract The chapter discusses the various consumer products manufactured out of black pepper and their market utility. The chapter will discuss, in detail, value addition in pepper, green pepper-based products, black and white pepper-based products, and other ancillary pepper-based products.

Keywords Value addition · Pepper-based products · Pepper-based by-products

Of all the spices used in the world, black pepper is, perhaps, the most widely used in the kitchen, in perfumery, in medicine, and in industry. Most consumer products that have their origin in black pepper are "value-added." For instance, when pepper is used in the food-processing industry, it is not pepper as such, but the oleoresin extracted through solvent extraction, and its pungency and flavor are the elements that add to the value of the end product made by the food-processing industry.

K. P. Nair, *The Geography of Black Pepper (*Piper nigrum*)*,
https://doi.org/10.1007/978-3-030-52865-2_13

Chapter 14
Value Addition in Pepper

Abstract The chapter discusses, at length, the various pepper-based value-added products, green pepper, etc. These products have tremendous market potential.

Keywords Black pepper · Green pepper · Ancillary pepper-based by-products

There has been a dramatic advancement in the field of value addition of black pepper and diversification of processed pepper products. The value-added pepper products are classified as follows:

1. Green pepper-based products
2. Black and white pepper-based products
3. Pepper by-products

These products may be further classified, as discussed in the following sections (Pruthi 1997).

14.1 Green Pepper-Based Products

1. Bottled green pepper in brine
2. Canned green pepper in brine
3. Bulk-packed green pepper in brine
4. Cured green pepper (without any covering tissue)
5. Freeze-dried green pepper
6. Frozen green pepper
7. Green pepper pickles in oil, vinegar, or brine
8. Green pepper-mixed pickle in oil, vinegar, or brine
9. Green pepper-flavored products
10. Green pepper paste
11. Sun-dried or dehydrated green pepper

K. P. Nair, *The Geography of Black Pepper (*Piper nigrum*)*, https://doi.org/10.1007/978-3-030-52865-2_14

Pepper fruits that are not fully mature are used in the manufacture of green pepper-based products, such as pepper in brine, pepper in oil, pepper in vinegar, desiccated green pepper, freeze-dried green pepper, and pepper paste. Almost all green pepper products are used by the catering sector to be served with meat dishes, such as steak and pork. The food industry also uses green pepper in the production of certain specialized cheeses. The other green pepper-based products are pickled green pepper and pepper spike. Tender pepper spikes and fruits alone or in combination with tender cardamom, pickled in vinegar and salt or sugar, make very delicious dishes that can be served with dinner. Green pepper paste is another product and is now a common sight in supermarkets, in polypacks or bottles. It can be used as a substitute to pepper powder, and it gives a refreshing taste with flavor and a unique "bite," which enhances the palatability of fish and meat dishes. It is a common choice of well-trained chefs.

14.2 Black and White Pepper-Based Products

1. Black pepper powder
2. White pepper powder
3. White pepper whole
4. Pepper oleoresin
5. Pepper oil
6. Other pepper products
7. By-products from pepper waste
8. Miscellaneous forms used in medicine, culinary, and industrial applications

The Pepper Marketing Board of Malaysia (PMBM), as well as many privately owned industrial houses, have brought to the market a number of innovative end products of pepper and have facilitated planning and product development of black pepper. The following structure has been proposed by the PMBM (Abdulla 1997). The pepper products are whole black pepper in retail packs for table use and ground pepper in retail packs and dispensers. Pepper-based products are flavored ground pepper, such as lemon pepper and garlic pepper; sauces; marinades, which have pepper as the primary component; and pepper paste. Spice mixtures and blends are curry powders and spice blends for specific cuisines, such as five-spice powder and soup blends. In addition, there are pepper-flavored products, which are pepper mayonnaise, pepper tofu, pepper cookies, and pepper keropok (which are prawn or fish crackers). Products that use pepper extracts are pepper candy, pepper perfume, and the like. Pepper is also put to other auxiliary uses, such as in the case of pepper stalk, which is used as a substrate for growing mushrooms.

There are a number of common black pepper products, of which the most important is ground pepper. This is the exclusive form in which black pepper is used on the table. It is also the only spice served along with food (in flight) on board and used in restaurants specializing in fast foods. Pepper is very commonly used for seasoning food at different stages of preparation. The following factors have to be kept in mind in the case of ground pepper use, which is the easiest to manufacture and market: (1) Freedom from bacteria and mold is extremely important; otherwise, there is a potential health hazard. (2) During the grinding process, volatile oil content can be adversely affected. Care must be taken to ensure that this does not happen. (3) Because high moisture content in the berries will adversely affect shelf life, moisture during storage should be kept to a minimum. (4) Optimum particle size should be ensured for free flow during the period of storage and active use. (5) Packaging is also very important. Without good packaging, the product will not have a good market value and could deteriorate during storage. (6) Above all, marketable pepper should be free from all extraneous matter.

Flavored pepper is an important product with a good market for culinary purposes. Lemon and garlic pepper are examples. The former contains lime powder, pepper powder, and salt, and the latter is a blend of dehydrated garlic powder and dried pepper powder. Both are table condiments, especially in nonvegetarian dishes, containing chicken and fish. The PMBM has developed these products. Pepper sauce is another product of PMBM that can be used as a marinade, especially in meat-based dishes such as steak and lamb; it is also used in stir-frying. The preparation of this product involves the use of black pepper, soybean extract, garlic, and other ingredients to obtain a relatively mild dark sauce. Pepper is also used extensively in the preparation of a variety of sauces including Worcestershire sauce.

Spice mixtures and blends are extensively used in continental kitchens as well as in regional kitchens. Pepper-based spice mixes, available in departmental stores and supermarkets, are widely used in different meat-based preparations. There are also soup mixes that contain pepper. Private industrial houses and PMBM have developed a variety of pepper-flavored products, such as pepper-flavored mayonnaise, egg tofu, savory pepper cookies, traditional biscuits with pepper flavor, and pepper-flavored prawn and fish crackers. When dressed with pepper, many traditional dishes and snacks will result in enhanced taste and flavor. Pepper-flavored vinegar and pepper salts are excellent taste enhancers.

Pepper is widely used in industrially manufactured food items. Almost all fish- and meat-based products use pepper for seasoning, which enhances both flavor and pungency. Industrially manufactured foods, such as soups and pickles, use pepper extracts or powdered pepper. Additionally, there are pepper-flavored beverages such as pepper tea, pepper coffee, pepper-flavored milk, and the like. Some brands of brandy contain piperine and pepper oil, which impart a pungent taste and an "exotic" flavor. In the perfumery industry, volatile oil from pepper is used to impart an exclusive spicy "oriental" touch. Some of the more popular brands of perfume that contain pepper volatile oil are Revlon's "Charlie" and Christian Dior's "Poison" and Malaysian perfumes, such as "Sensai" and "Amila" (Ng 1993).

14.3 Other Ancillary Pepper-Based By-Products

An industrial area worth greater research and development is that of ancillary pepper-based by-products, which are of uncommon use. For instance, Malaysia has succeeded in producing paper and board from pepper stalks. These materials—especially paper made of pepper stalk—make good invitation cards, having a unique texture and color pattern (Ng 1993). Pepper stalk makes a good substrate for oyster cultivation when grown in a 1:1 mixture of pepper stalk and shredded paper (Siti Hajijah and Bong 1993). In Malaysia, the remnants of the pepper-processing industry, which consists mainly of stalk and pericarp, are powdered and used as an organic manure called pepper dust. It can enrich soil fertility when used in combination with other organic agricultural wastes (Ng 1993).

In addition, there are a number of other pepper-based commercial formulations. Though a number of commercial formulations, such as canned green pepper, green pepper in brine, pepper oil, and oleoresin, are readily available in the market, they have yet to find widespread acceptability by the consumer. The most recent use of spice-processing technology is encapsulation of the flavoring components—namely, spice oils and oleoresin. These can be fully exploited commercially on a large scale, provided capital and entrepreneurship are readily available. This is yet to happen.

There is now a changing consumer preference for white pepper from black pepper in some Western countries. White pepper is used both as table pepper and also for enriching the flavor and taste of crab soups, seafood salads, casseroles, chicken, fish, and egg preparations. Sauces like mayonnaise also contain white pepper as an ingredient. One of the primary reasons for this shift in preference is because the black specks from the skin of black pepper are unseemly to the eye on a dining table.

The pepper product that is becoming increasingly popular in European countries is the green pepper. Because the piperine and pepper oil levels are high in slightly immature corns, they are preferred in the preparation of green pepper. This gives a characteristic flavor and is soft with a "bite"—a unique feel on the palate. These qualities make it ideal for garnishing meat dishes. The major products are green pepper in brine, canned green pepper, dehydrated green pepper, freeze-dried green pepper, and green pepper paste. While an increasing number of consumers prefer green pepper in brine and green pepper paste and while demand for such is on the rise, demand for canned pepper is not high because its cost of production is quite prohibitive. India, Brazil, and Malaysia are the major producers of these value-added pepper products. The catering industry is the main source of demand for these products, followed by the food manufacturing industry.

Among the flavoring agents, the most important are oleoresin, pepper oil, and encapsulated pepper. The most extensive use of oleoresin is in the flavoring of meat. The other end uses of oleoresin are for preparing pickles, sauces, gravies, dressing *chutneys* (unique Indian spicy eatables), soups, and snacks. In most meat preparations, pepper is used to impart flavor. The same task is also accomplished by oleoresin. One of the most recent advances made by the Regional Research Laboratory in Trivandrum, Kerala State, India, is in developing the spray-drying encapsulation technique for oleoresin. The encapsulated flavor powder is used in a variety of products, such as cake mixes, dry beverage mixes, desserts, soup mixes, dusting on potato chips, and nuts. These have an emerging market all over the world.

References

Abdulla A (1997) Product development and promotion of pepper products. In: Proceedings of the 22nd Pepper Tech. Meeting. Spices Board, Cochin, Kerala State, India, pp 1–6

Ng SC (1993) Uses of pepper. In: Ibrahim MY, Bong CFJ, Ipor IB (eds) The pepper industry: problems and prospects. Universiti Pertanian Malaysia, Sarawak, pp 347–353

Pruthi JS (1997) Diversification in pepper utilization. Part I & II. Green pepper products. Int Pepper News Bull 14(4): 5–9; 14(5):6–9

Siti Hajijah AS, Bong CFJ (1993) Pepper fruit stalk as substrate for mushroom cultivation. In: Ibrahim MY, Bong CFJ, Ipor IB (eds) The pepper industry: problems and prospects. Universiti Pertanian Malaysia, Sarawak, pp 342–346

Chapter 15
Conclusions and a Peep into Pepper's Future

Abstract The chapter discusses at length a peep into the future of pepper on a global scale. There are several dimensions to this topic; possibly, the area that has much promise is the development of pepper-derived food additives. The potential of this is gradually unfolding on an industrial scale. The chapter discusses these aspects.

Keywords Pepper · Future · Global economy · Pepper-derived food additives

Pepper has certainly had a checkered history. Within the past, present, and future, pepper can indeed be termed the "King of Spices." Though beset by many problems, both economic and agronomic, it is a safe bet that pepper will emerge as the world's most sought-after spice. It has been estimated that by the year 2020, global demand for pepper will be about 280,000 metric tons, which is projected to escalate to 360,000 metric tons by the year 2050. This would entail almost doubling the current production in the first half-century of this millennium. Where will the additional output come from? Increase of the pepper-growing area could only be marginal on a global basis. India, the major producer, will, perhaps, experience only marginal area increases. On the Asian subcontinent, it is only in China and Vietnam (the latter already emerging as a key producer) that marginal area increases can be expected. Like India, only a marginal increase in area can be expected in Indonesia. Malaysia and Thailand, the other two pepper producers of Asia, are already experiencing a decrease in area. This is also true for Brazil, a key producer on the Latin American continent. Labor is becoming increasingly difficult to obtain and very expensive when found; hence, countries such as Malaysia and Thailand on the Asian continent and Brazil on the Latin American continent are moving away from the labor-intensive pepper production. The impact of globalization is clearly seen in these countries, where market forces are shying away from labor-intensive crops such as pepper in preference to more mechanically managed crops. By the very nature of the pepper plant and the pepper canopy, mechanization has only very limited possibilities.

K. P. Nair, *The Geography of Black Pepper (*Piper nigrum*)*,
https://doi.org/10.1007/978-3-030-52865-2_15

From the foregoing review, it can clearly be seen that one of the most pressing needs of pepper production is to have a pepper ideotype that combines many positive attributes in terms of boosting production potential and, at the same time, resisting many ravages of nature—especially those of diseases, among which the most predominant is foot rot, caused by the *Phytophthora* fungi. There is yet no single pepper variety that is universally resistant to this dreaded disease, though some lines have been identified that have shown regional promise. Conventional breeding, which will take years to accomplish because pepper is a long-duration perennial, will not be the answer to this major problem. Biotechnology is a promising area, but there are vast gaps in knowledge. Behavior of the pepper plant at the molecular level is far from clearly understood. Though tissue culture is a promising area in terms of cutting down the long gap in producing a desirable ideotype, this will not give final answers because there are far too many gray-area variables in the technique. Gene sequencing, currently a very highly sophisticated technology with very high levels of investment, opens up a door in the case of pepper as well. At the current level of global scientific knowledge and available technical expertise in the case of pepper, it is a safe bet to say that among those who exhibit a great fascination with crops, especially pepper, it is a very long road to traverse. Sophisticated instrumentation is only part of the answer; the availability of superlative technical skills is of great importance, but the pepper-producing countries, India included, clearly lack such skill. What one might expect on the pepper front, at least for the next few decades, is the following.

As science advances and global populations become more and more health conscious, the demand for organically grown food will become greater. One of the most promising areas for future pepper development will be in the area of producing "organically grown" or what one might term "clean" pepper. The environmental scare generated by the so-called green revolution—the hallmark of which has been extensive and indiscriminate use of chemical fertilizers and pesticides—is driving more and more people to organically grown food, and the demand for organically grown pepper is no exception. The potential market for organically grown food products, including pepper, is most promising in the USA, followed by Germany. The rigors of growing organic foods are encompassed in the Codex Alimentarius stipulations developed by France, and the USA has its own stipulations. In a country like India, where pepper production on the farm and its further processing is not subject to the rigors of quality control, export of organically grown pepper can only be put on sound footing if production and processing methods are well streamlined and rigorously controlled for quality, sanitation being the most important. The same would hold for the situations in other pepper-producing countries like Indonesia, Vietnam, Malaysia, and Thailand.

Another area that has much promise is the development of pepper-derived food additives. The potential of this is gradually unfolding on an industrial scale. Heavy investments are necessary to make the various products price competitive. Some regional success stories do exist, though. For instance, there are pepper-based products that have a ready regional market. Yet, these products do not have a global reach. The potential market for these products would be Europe and the USA, where a gradual change in food habits is taking place. Both postharvest technology and product development would undergo remarkable changes in the future. Imaginative product diversification would lead to the development of an array of novel products for the consumer, currently unheard of. Pepper with high flavor but low pungency will begin to be increasingly used in the manufacture of liquors, fruit juices, bakery products, and even choice perfumes. A possible challenge to the development of pepper varieties with high flavor will be the creation of the in vitro pepper flavor from bioprocessors. If this comes about, the only way natural pepper can be salvaged is a preference of the consumer for the clean, natural spice, rather than the laboratory-created spice.

Another important area that needs to be thoroughly researched is management of the fertility of pepper soils. Almost the entire approach to fertility management in pepper soils is based on classical textbook knowledge, where empirical recommendations generated from micro-plots are extrapolated to large-scale field conditions. These experimental micro-plots are nothing but artifacts, and many of the recommendations that emanate from such studies, when applied to large-scale plantations, turn out to be quite off the mark in reproducing accountable results. This often shatters farmers' confidence. A very significant departure is being made after the concerted efforts of the author, who has developed an entirely new approach to soil testing and fertilizer management, based on nutrient buffering. The concept, developed by the author of this book, is now universally known as the "Nutrient Buffer Power Concept." A detailed discussion on the concept and its relevance in pepper nutrition is given in Chap. 4. In India, in the state of Kerala, where pepper is grown to the largest extent, it has been observed that fertilizer input can be significantly reduced by taking into consideration the buffer power of the nutrient concerned. Experimental and field results (Nair 2002) show that when compared with routine soil testing and fertilizer recommendations, addition of zinc fertilizer, a very crucial input in pepper production, can be reduced by almost 75% in some soils. Such breakthrough research escape the attention of agronomists and soil scientists steeped in routine textbook knowledge.

There is an urgent need to extend the concept to other crucial nutrients, such as phosphorus and potassium and possibly nitrogen as well. Of the first two, phosphorus is more important in view of the fact that soils of Kerala State, the home of pepper, are lateritic and much of the applied phosphate fertilizer, based on routine soil testing, is not available to the crops. It is in this context that this new concept holds much hope for pepper farming. However, it must be emphasized that the success of a new approach, to a great extent, rests with the ingenuity of those applying it to suit the demands of a new situation. This principle is no exception to making the nutrient buffer power concept succeed in the case of pepper production, as has been the case with other crops, such as maize, rye, white clover, and even a perennial crop like cardamom, tested by the author (Nair et al. 1997). The fact that pepper is a perennial crop makes it all the more important because unlike an annual one, where a midcourse correction can be effected in the following season, the fertilizer regime has to be corrected from the very beginning because pepper grows for upward of 25 years. Unlike routine soil testing, the new approach calls for an accurate determination of the buffer power of the nutrient in question at the very start. Once this is accomplished, the buffer power factor can be integrated into the computations with the routine soil test data, and accurate fertilizer recommendations can be made on the basis of this new information. This implies that in addition to obtaining routine soil test data, one also needs to know the buffer power. The author has obtained very encouraging results with the new concept in pepper production in Kerala with regard to zinc fertilization, which signals a very promising change in the fertilizer practices. Hopefully, the new concept could very successfully be extended to other important plant nutrients.

Agriculture is the engine for development, nationally and globally, as is being increasingly proved against the unfolding scenario of globalization. A nation with a strong agricultural base can be expected to be in the forefront of economic development. Pepper is the crop of the tropics and, in that sense, is very much a part of the development of the Third World economy. A lot has been achieved in pepper production, but before it moves to the center stage as a commanding crop of the Third World economy, much more needs to be done. This calls for the concerted efforts of not just agronomists, soil scientists, or plant breeders dealing with pepper but of a whole range of visionary thinkers and planners, who can really make the crop, not just the king of spices but the monarch of the Third World agricultural economy.

References

Nair KPP (2002) Sustaining crop production in the developing world through the Nutrient Buffer Power Concept. In: Proceedings of the 17th World Soil Science Congress, Bangkok, Thailand, August 14–21, vol 2, p 652

Nair KPP, Sadanandan AK, Hamza S, Jose A (1997) The importance of potassium buffer power in the growth and yield of cardamom. J Plant Nutr 20(7 & 8):987–997

Precise Quantification of Plant Nutrient Bioavailability for Black Pepper and Cardamom: The Relevance of the Nutrient Buffer Power Concept

Many years ago, in one of the early editions of *Advances in Agronomy*, the *magnum opus* of agricultural science, or what is popularly known as the "Bible of Agricultural Science," Roy W. Simonson writing a chapter entitled "Concept of Soil" noted: "Someone has said that the fabric of human life is woven on earthen looms – it everywhere smells of the clay." Almost a half century later, we agronomists and soil scientists have come very far in our understanding of the "fabric of human life" which "everywhere smells of the clay." That the "fabric of human life" which is so very intimately linked to soil has dramatically changed is beyond dispute. Yet, there is no denying the fact that the "fabric of human life" will always be linked to soil which is the "pragmatic, the reality, the entity that dictates much of what societies can do" (Boul 1994). Soil, in my opinion, is an invaluable gift of God to life on planet Earth and can aptly be termed "The Soul of Infinite Life." Though the basic concept of soil, since its early description as a "thin mantle over the land surface," has vastly changed over the years, this thin mantle has always been the focal point since it is the medium for plant growth. For early man, it was nothing more than a physical support for his predation. Quite likely, some areas were known to provide better footing than others, and some were to be avoided if possible. It is amazing that even after decades of research in soil science, which has provided such invaluable information on this "thin mantle over the land surface" so crucial to the existence of life, human, plant, and animal, on planet Earth, this basic instinct of predation has remained unchanged. How else can we explain the disdain and callousness so often witnessed in modern societies, propelled by an insatiable greed to acquire unlimited wealth, which leads to the abuse of soil, the invaluable gift of God to man? Undoubtedly, the earliest shift in attitude toward soils must have originated at a time when man began to grow food, rather than gather it as his ancestors did. In many ways, the shift in attitude was the precursor to modern-day soil science. Though this shift must have occurred in pre-Christian times, about 9000 years ago, and focused on the inevitability of a proportionally smaller land surface supporting larger human populations, it is only in recent times that we have

K. P. Nair, *The Geography of Black Pepper (*Piper nigrum*)*,
https://doi.org/10.1007/978-3-030-52865-2

witnessed the magnitude of the impact of this shift in attitude of human existence. Much of land has become degraded and unsuitable for agriculture since a century ago. The 1992–1993 World Resources Report (Stammer 1992) from the United Nations on the status of world soils contains very alarming conclusions. For example, nearly 10 million ha of the best farmlands of the world have been so ruined by human activity since World War II that it is impossible to reclaim them. Over 1.2 billon ha have been seriously damaged and can be restored only at great cost. The loss in soil capability could mean that there will be enormous food shortages in the next 20–30 years and, as is but natural, the people of disadvantaged nations will suffer the most. Many factors have continued to this alarming state of affairs, one of the prime factors being the "high-input agriculture" or more specifically the "chemical agriculture," euphemistically called the "green revolution," where unbridled use of chemical fertilizers coupled with carcinogenic herbicides and pesticides has ruined the soil resources. South Asia, in particular India, is a prime example. The state of Punjab known as the "cradle of green revolution" is a living example where hundreds of acres of once fertile land has turned barren, where even a blade of grass will not grow now anymore. The groundwater is so polluted with nitrogenous residues from excess of urea, and water is no more potable. Nair (2019) has reported that the excess use of urea has led even to global warming, because of the loading of stratosphere with nitrous oxide (N_2O), a by-product of urea hydrolysis—a fact hitherto hidden by vested interests of the fertilizer industry. Dried aquifers, and vanishing biodiversity (Nair 2019) due to continuous monoculture of wheat and rice, are environmental hazards. There are hundreds of acres of land stretches that have become barren, where once stood lush wheat and rice fields. Crop yields have plateaued or drastically declined. Two-thirds of the seriously eroded land is in Asia and Africa. About 25% of the cropped land in Central America is moderately to severely damaged. In North America, this is a small percentage—only 4.4%. Since the time of the green revolution, food production has declined dramatically in 80 developing countries during the past decade. Soil degradation is the major factor. In India, of the 328.73 mha of geographical area, as much as 120.40 mha (nearly a third) has degraded soils, all due to the after effect of green revolution. Nearly 40% of the world's farming is done on very small parcels of 1 ha or less (Robinson et al. 1981). Ignorance and poverty characterize this situation. Yet, emphasis on agriculture has always been confined mostly to large-scale farming. Large-scale farming, grand projects at huge costs and huge profits, has been the order for many decades. In a lighter vein, it can be said that even the "lebens raum" concept of Adolf Hitler, the master military strategist, who set out to conquer the world, had an echo in the inevitability of this modern-day fact. What else can justify the ruthless conquest of vast territories of land by the master military strategist, who set out to conquer the world or, more appropriately, the world's soils?

Despite the complexity of soil science and the emergent soil management practices, the basic concept of soil as a medium of plant growth can be expected to persist for an indefinite length of time. But it is becoming increasingly clear that the earlier views on soil as merely the "supportive medium" for plant growth is giving place to newer ones on "managerial concepts" of this supportive medium. This is amply illustrated by the shift in focus from the green revolution phase of the 1960s to the mid-1970s, where application of increasing quantities of soil inputs, such as chemical fertilizers and pesticides, was emphasized, to the "sustainable agriculture" phase from the early 1980s to the present (probably to continue?). Sustainable agriculture places more reliance on biological processes by adopting genotypes to adverse soil conditions, enhancing soil biological activity, and optimizing nutrient cycling to minimize external inputs, such as fertilizers, and maximize their efficiency of use. In fact, the paradigm of the earlier phase has given way to the emergent new paradigm (Sanchez 1994), and this is clearly reflected in the dialogue of the world leaders during the Earth Summit in 1991 in Rio de Janeiro, Brazil, where Agenda 21 has incorporated six chapters on soil management issue (Keating 1993). The focus of this review will be on the second paradigm inasmuch as prescriptive soil management is concerned with regarding to understanding the dynamics of soil nutrient bioavailability, with specific focus on "The Nutrient Buffer Power Concept."

Efficient Plant Nutrient Management: The Key Factor in Sustainable Black Pepper Production

Agricultural systems differ from natural systems in one fundamental aspect: while there is a net outflow of nutrients by crop harvests from soils in the first, there is no such thing in the second (Sanchez 1994). This is because nutrient losses due to physical effects of soil and water erosion are continually replenished by weathering of primary minerals or atmospheric deposition. Hence, the crucial element of sustainability of crop production is the nutrient factor. But of all the factors, the nutrient factor is the least resilient (Fresco and Kroonenberg 1992). The thrust of "high-input technology," the hallmark of the green revolution, in retrospect, and the moderation of the "low-input technology," the foundation stone of sustainable agriculture, in prospect, both dwell on this least-resilient nutrient factor. If the pool of the nutrients in the soil, both native and added, could be considered, the "capital, efficient nutrient management might be analogous to raising the interest" accrued from this capital in such a way that there is no great danger of the erosion of this capital. Hence, sound prescriptive soil management should aim at understanding the actual link between the "capital" and the "interest" so that meaningful management practices can be prescribed.

Soil Tests and Nutrient "Bioavailability"

It is universal knowledge that soil tests are the basis for predicting nutrient "bioavailability." There are, perhaps, as many soil tests for each nutrient as there are nutrients. This review will not dwell on the merits or demerits of any single soil test

Table 1 Nutrient balance (kg/ha/year) in intensively managed arable soils

	N	P	K
Inputs			
Fertilizers	156	39	119
Other	32	–	9
Total	188	39	128
Outputs			
Harvest	103	16	91
Removal	24	–	–
Other	–	–	–
Total	127	16	91
Balance	61	23	37

Note: After Frissel (1978)

or a group of them. Suffice to say that fertilizer recommendations traditionally are made at the point where marginal revenues equal marginal costs, which involve some positive synergism (De Wit 1992). The most common result of this approach is the vast buildup in the soil nutrient pool in intensively cultivated soils (Whitmore and van Noordwijk 1994). Data in the Table 1 indicate positive balances (in kg/ha/year) for N (61), P(213), and K(37) in intensive crop production systems (Frissel 1978).

Over several decades, such positive balances can lead to a huge buildup of the nutrient capital, especially in the case of high-input, intensive agricultural systems, as in the case of many European, North American, and Scandinavian countries. A dangerous consequence of such huge soil buildup is nutrient contamination of groundwater to such extremes that "environmental soil tests" become necessary to assess critical limits of nutrient pollution (Sharpley et al. 1993). In a state like Punjab in India, the "cradle" of Indian green revolution, groundwater has been rendered nonpotable (Nair 2013). Nitrogen is a prime candidate for this scenario, especially in the temperate zone. At the other end of the spectrum are the marginal areas of tropical zone where inadequate replenishment of nutrient removal by crop growth and also nutrient loss by soil and water erosion has left that capital "in the red." Initially fertile alfisols of much of Africa with subsequent severe depletion of N and P (Yates and Kiss 1992) are an example of this "nutrient bankruptcy." Either way, contemporary soil tests are the basis on which prescriptive management practices are formulated.

Rating Soil Tests to Define Nutrient Bioavailability and a Fertility Index

Most soil test laboratories around the world use some kind of "rating system" to evaluate soil test values. These rating systems invariably use qualitative terms, such as "low," "medium," and "high," to describe the bioavailability of a specific soil nutrient. Admittedly, these terms denote different meanings in the context of soil nutrient bioavailability of a particular plant nutrient and, at best, are empirical technologies. This problem has been recognized by researchers over the years. Morgan (1935) suggested a scale of 1–10 with 8 equal to the point of no response. Bray and Kurtz (1945) used relative yield or percentage sufficiency to describe the deficiency, with 100 defined as the point of no response. The index below 100 follows the curvilinear relationship between soil test values and yield without the addition of the element. Above 100, the index displays a straight-line relationship indicating the relative margin of adequacy or the proximity to an excessive level. To eliminate the need for a percent sign, the values are referred to as "fertility indices," and they are reported to the nearest multiple of 10 from 0 to 9990 (Cope and Evans 1985). In addition to ratings, most laboratories use some method of reporting results more precisely, mainly for use by farmers in record keeping and monitoring soil fertility. Some report kg/ha or lb/a or ppm extracted, but these would be confusing to farmers, because each element has a different level for a specific degree of adequacy (Cope and Evans 1985). For instance, the adequate or critical level for one soil may be 25 ppm P, 120 ppm K, 200 ppm Ca, and 30 ppm Mg. Adequate levels in other soils and from other extracting procedures would be different for each element (Cope and Evans 1985).

Despite the fact that a number of soil tests and others such as Diagnosis and Recommendation Integrated System (DRIS) are in vogue to predict nutrient bioavailability, it must be said that a universal picture is yet to emerge in this field with regard to precise availability prediction. This is primarily because a soil test in the laboratory can never simulate plant root absorption of a nutrient in a field soil, though most of the time the assumption is that it does. In the final analysis, it is the plant and plant alone which will decide whether or not the nutrient is bioavailable. This report examines the important question whether there is an alternative and more precise approach to quantify the nutrient bioavailability. And it is in this context that "The Nutrient Buffer Power Concept" comes under focus.

Experimental Evidence in Black Pepper

The theoretical background for the concept is explicitly explained in the chapter titled "The Buffering Power of Plant Nutrients and Effects on Availability" by Nair (1996) in *Advances in Agronomy*, Vol. 57, 237–287. The field experimental results are discussed below.

The following discussion would pertain only to zinc, the most important micronutrient in black pepper cultivation, with some reference to wheat as well, from Central Asia, for the sake of comparison.

Quantifying the Zn Buffer Power of Pepper-Growing Soils

Nair (1984) has used electro-ultrafiltration (EUF) technique to measure the Zn "intensity" and an adsorption-desorption equilibrium technique to measure the Zn "quantity." Because the EUF technique is a highly sophisticated instrumental procedure, where an expensive electro-ultrafiltration instrument has to be set up, only rich European and North American countries can afford to employ such techniques. For the developing countries, a simple adsorption-desorption equilibrium technique has been used as follows (Nair 2002):

200g of representative soil samples from the principal black pepper-growing regions in the State of Kerala, India, were incubated at 60% of the maximum water-holding capacity with graded rates of Zn over a fortnight, by maintaining the water regime to constancy through watering each day at the same time of the day. At the end of the period, the soil samples were extracted with 0.01 M $CaCl_2$ solution over a 24-hour period and the extracts tested for Z using atomic absorption spectrometry. This represented the Zn "intensity." Separate extractions were made with DTPAS, which represented the Zn "quantity." Table 2 show the Zn buffer power of the experimental soils.

Discussion

Results in Tables 2, 3, and 4 unequivocally establish the importance of Zn buffer power integration in the computations in the predictability black pepper plant top Zn concentration, plant top Zn uptake, and, eventually, dry matter production in black pepper. DTPA (diethylenetriaminepentaacetic acid) is the universally adopted soil test for Zn bioavailability. Most remarkably, the overall relationship between DTPA soil test and black pepper dry matter production not only improved very significantly, when the Zn buffer power was integrated into the computations, but changed from a negative relationship to a positive one with the Zn buffer power integration into the computations (Table 4). This is a very remarkable scientific finding emphasizing the importance of "The Nutrient Buffer Power Concept." Further, it also

Table 2 Zinc buffer of the black pepper-growing soils from India

Soil	"r" value	"b" value
Peruvannamuzhi	0.8337***	0.7824
Thamarassery	0.9304***	1.5786
Ambalavayal	0. 9604***	3.0358

Note: All the "r" values are very highly significant, at a confidence level of 99.9% (***), implying that the adsorption-desorption equilibrium technique developed by the author provides an accurate measure of the Zn buffer power of the concerned soils. Further, the soils showed widely varying Zn buffer power, as shown by the "b" values, with the soils from *Peruvannamuzhi* region least buffered for Zn, while that from the *Ambalavayal* region most buffered, with the soils from *Thamarassery* region intermediate in their Zn buffer power. These findings have remarkable impact on black pepper production, as we shall see from results incorporated in Tables 3, 4, and 5

Table 3 Linear regression functions (Y = a + bx) and correlation coefficients ("r" values") for the relationship between DTPA extractable Zn (Zn "quantity." Y) and 0.01 M$CaCl_2$ extractable Zn (Zn "intensity," x)

Soil	Adsorption-desorption equilibrium	Soil incubation	"r" values	
			A	B
Konya (Central Anatolia)	$Q = 201bI + 5.3$	$Q = 679bI$-5.9	0.951*	0.978*
Koruklu (Gap region)	$Q = 17bI + 17.9$	$Q = 375bI -$ 8.7	0.015	0.784*
German soil (reference soil)	$Q = 8bI + 6.5$	$Q = 11bI + 7.8$	0.972*	0.942*

Note: Q = Zn quantity, I = Zn intensity, b = Zn buffer power. A, without the integration of Zn buffer power ("b" values) in the computations; B, with the integration. The experimental technique employed has resulted in very highly significant correlation coefficients ("r" values), except in the case of soils from the Gap region (Koruklu) in the case of the incubation experiments, which suggest the high accuracy of the adsorption-desorption equilibrium technique to determine Z intensity, which is a true reflection of the Zn buffer power. Note the remarkable improvements in the "r" values (B compared to A) which unequivocally show that the integration of the Zn buffer power remarkably improves the relationship. How this affects plant uptake, and eventually wheat yields, will be discussed in the following tables and explanation of experimental details
*Significant at a confidence level of 99.9%

Table 4 Black pepper yields from farmers' fields—the figures are weighted against the respective Zn buffer power of the representative soils

Soil	Yield (kg/vine)		Deviation (%)
	Targeted	Actual	
Peruvannamuzhi	0.241	0.401	+66
Thamarassery	0.490	0.487	+0.6

Note: Target weighting was done against the highest vine yield obtained from the *Ambalavayal* region. Note the remarkable closeness between targeted and actual yields in *Thamarassery* region, whereas in *Peruvannamuzhi* region, there is a huge deviation. Soils in the *Peruvannamuzhi* region are atypical black pepper soils. Ironically, the "official" recommendation for Zn fertilizer application emanates from experiments conducted in *Peruvannamuzhi* region, as the black pepper research station, affiliated to the Indian Institute of Spices Research, Calicut, Kerala State, affiliated to the Indian Council of Agricultural Research, New Delhi, is located here

Table 5 The 'r" values (correlation coefficients) for the relationship between routine DTPA soil test for Zn bioavailability, Zn uptake, and dry matter production in black pepper with (A) and without (B) Zn buffer power integration

Details of the soil test	A	B
DTPA soil test vs. Zn concentration in plant top	0.8*	0.9*
DTPA soil test vs. Zn uptake by black pepper plant	0.8*	0.9*
DTPA soil test vs. black pepper dry matter production	−0.7	0.8*

Note: *Significant at a confidence level of 99.9%

demonstrates that the predictability of black pepper yield can be remarkably improved by taking into consideration the Zn buffer power (Table 4). This finding will immensely help the black pepper farmers of the State of Kerala.

Table 6 Effect of Zn fertilization (100 kg ZnSO4/ha) on mean and range of Zn concentration and wheat grain yield among 28 wheat genotypes grown in Konya soils and 25 grown in Gap soils in Turkey

	Tissue concentration (ppm)				Grain yield (kg/ha)				
	Without Zn		With Zn		Without Zn		With Zn		Percentage
Soil	Range	Mean	Range	Mean	Range	Mean	Range	Mean	increase
Konya	7–10	9	10–19	14	1550–4010	2710	248–5900	4570	69
Gap	16–23	19	22–31	25	1530–5260	3500	1550–5300	3510	0

Table 7 Interrelationship between Zn bioavailability parameters and wheat grain yield and shoot tissue concentration of Zn in Turkey

Parameter investigated	Correlation coefficient ("r" values)
DTPA soil test vs. Zn concentration in shoot tissue	0.672***
Zn buffer power versus percentage increase in wheat grain yield	0.860***

Notes: Note the very highly significant (at a confidence level of 9.99%***) "r" values between Zn buffer power and percentage increase in wheat grain yield. Though DTPA soil test seems to have a positive influence on Zn concentration in plant top, it is the Zn buffer power that yields far greater influence on wheat grain yield

As a reference investigation, the following one was conducted in Central Asia (Turkey) to examine the crucial question whether "The Nutrient Buffer Power Concept" holds the key to optimal Zn fertilization in wheat. Turkey is one of the major wheat-producing countries in the world. Nearly 45% (4.5 million ha) of the wheat-producing area in Turkey is located in the highly calcareous region of Central Anatolia. Central Anatolia is semiarid, being the driest region of Turkey. Scanty rainfall (only about 325mm) and low water supply greatly limit wheat yields. Zn bioavailability (as per the common DTPA test) has been found to be very low, and the farmers have been advised to apply as much as 100 kg of Zn fertilizer/ha, $ZnSo_4$.

Very recently, Nair (2019, unpublished data) conducted an investigation into the problem, and the results are summarized in Tables 6 and 7.

Discussion

The Zn buffer power is very clearly reflected in grain yield increase. Note that the "b" value (Zn buffer power, Table 3) for both the adsorption-desorption equilibrium technique (developed by Nair 2013) and incubation technique for Konya soils are 201 and 679, respectively. The corresponding values for Koruklu soils (Gap region) are 17 and 375, respectively. The percentage increase in wheat grain yield in Konya soils is 69 while that in Koruklu soils is 0. In other words, the Konya soils are highly Zn-buffered, and, hence, the Zn bioavailability to wheat plant is very high, which has reflected in high wheat yield.

Experimental Evidence in Cardamom

Cardamom is a commercial crop, and it has a great commercial significance for a country like India. In India, it is mainly grown in the States of Kerala and Karnataka. The crop is also grown in Guatemala, but the Indian cardamom is the world's choice. The main problem in cardamom cultivation is one of precisely quantifying its K fertilizer needs, as the crop is a very heavy feeder of potassium. The recent decontrol of prices of both potassic and phosphatic fertilizers in India has put the cardamom farmers in jeopardy, and it was thought necessary to explore whether "The Nutrient Buffer Power Concept" has any significant relevance in precisely quantifying the potassium fertilizer needs of cardamom. As of now, the routine NH_4OAC (ammonium acetate) test is used for formulating a potassic fertilizer requirement for cardamom. Hence, an extensive investigation was taken up in the States of both Kerala and Karnataka to obtain soil samples and leaf samples and examine how the potassium (K) buffer power determines the cardamom uptake and consequently cardamom yield. The experimental techniques are discussed elsewhere (Nair 2013). The salient results are incorporated in Tables 8, 9, and 10.

Table 8 The K buffer power of cardamom-growing soils

Region	Regression function (Y = a + bx)	Correlation coefficient
Coorg (Karnataka State)	Y = 142.38 + 1.443x	r = 0.856*
Idukki (Kerala State)	Y = 592.5 + 0.917x	r = 0.580**

Notes: "b" values in the regression functions represent the representative K buffer power of each soil. The K buffer power refers to the pooled values of soil samples obtained from 94 locations covering an area of more than 20000ha in the cardamom-growing regions of India
*Significant at 95% confidence level, **significant at 99% confidence level

Table 9 The relationship between leaf potassium and exchangeable potassium

Details	Regression function	"r" value
*Potassium in leaf vs. exchangeable potassium	Y=1.27 +0.0004	0.206
**Potassium in leaf vs. exchangeable potassium	Y= 1.65+0001	-0.006

Note: *Coorg soils, **Idukki soils

Table 10 Relationship between potassium in leaf and exchangeable potassium in soil for the pooled data of 94 locations for both soil regions without (A) and with (B) buffer power integration in the computations

	"r" values	
Details	A	B
Potassium in leaf vs. exchangeable potassium	0.25	0. 44**

Note: **Significant at a confidence level of 99%

Discussion

The experimental results show that the Coorg region had a very high yield of 155 kg/ha as compared to that from Idukki region at 80kg/ha—almost a 100% increase. This is simply due to the better K buffered soils of Coorg compared to that of Idukki region reinforcing the validity of "The Nutrient Buffer Power Concept." Results in Tables 8, 9, and 10 substantiate this conclusion.

Innovations in Spice Production

India as a traditional producer, consumer, and exporter of spices in the world holds a major stake in the global spice market with an estimated production of 8202 thousand tons of spices during 2016–2017 from 3705 thousand ha (Fig. 1). Though every state in the country grows at least a few spice crops, Kerala, Andhra Pradesh, Gujarat, Maharashtra, West Bengal, Karnataka, Tamil Nadu, Orissa and Madhya Pradesh, Rajasthan, and North Eastern states are the major spice-producing states. The increase in spices production from 2015–2016 to 2016–2017 was the highest (17.4%) among horticulture crops production. In a total of 43 principal agricultural commodities exported from India, spices stand fourth after marine products, buffalo meat, and basmati rice. Indian spices are exported to more than 100 countries; however, main importers are the USA, China, Southeast Asian countries, the UAE, the UK, Germany, Saudi Arabia, Thailand, the Netherlands, Sri Lanka,

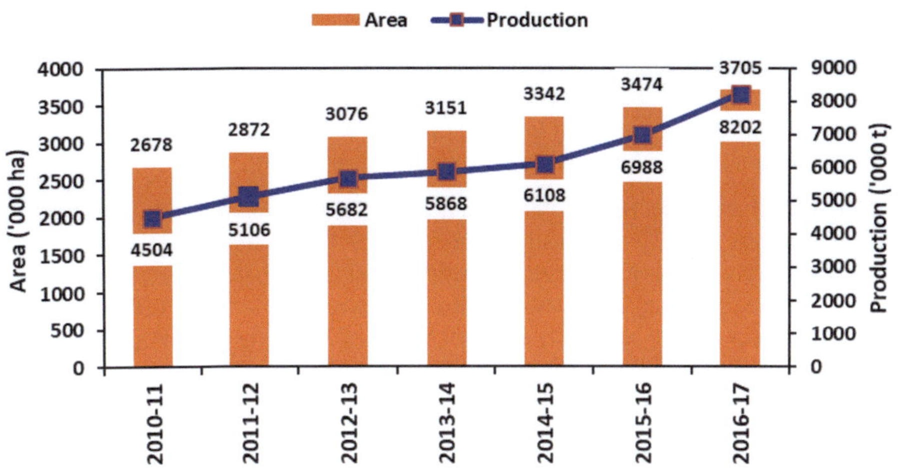

Fig. 1 Trends in spice area and production

K. P. Nair, *The Geography of Black Pepper (*Piper nigrum*)*,
https://doi.org/10.1007/978-3-030-52865-2

Mexico, Bangladesh, Brazil, Pakistan, Japan, France, Egypt (ARE), Spain, South Africa, and Australia.

In terms of export, a total of 10.07 lakh tons of spices and spice products valued at Rs. 19367.00 crore have been exported from the country during 2016–2017 contributing 8.70% to the total agricultural product export. It is noteworthy to mention that only around 10.00% of our spice production is exported and it commands 43.00% share in volume and 48.00% in value of the world spice trade. The remaining spices are consumed in India which is an indication of the prevalence of the huge domestic market within the country. Among the export of different spices, maximum share was from chile followed by seed spices, turmeric, and black pepper. However, in terms of value, mint products and spice oils and oleoresins contributed a major share of the total export earnings.

The food and pharmaceutical industry rely on these high-value low-volume commodities to produce an assortment of products that are designed to meet the varied needs of the consumers across the world. The estimated growth rate for spice demand in the world is around 3.19%, which is a shade above the population growth rate.

Major Challenges

The major challenges confronted in the spice production sector are:

- Climate change resulting in drought/excess moisture, high/low temperature during critical periods, etc.
- Emergence and epidemics of pests and diseases
- Adulteration of spices
- Pesticide residues and mycotoxin contaminants in the products and lack of MRL and ADI standards in some of the pesticides used in spices
- Shifting of interests of growers to more profitable/less risky crops
- Cyclic market fluctuations at international and national levels
- Competition from other major spice-producing countries

In view of these challenges, there is a need to develop cutting-edge technologies that are simple, cost-effective, and farmer-friendly. Apparently, the spice sector has to inadvertently deal with many other risky situations, where the possibilities for adverse consequences that can cause marked disruptions are very high. There is great scope to improve the productivity of major spices by adopting technologies that will help to bridge the gap between potential yields realized in the research stations/ progressive farmers' plots. ICAR-Indian Institute of Spices Research (IISR) and All India Coordinated Research Project on Spices (AICRPS) are working together on mandated spice crops production and productivity enhancement and its utilization through research advancements in developing new varieties, plant health management technologies, and value addition.

Technological Innovations

1. **Improved Varieties of Spices**

ICAR-IISR has released several high-yielding and improved varieties of ginger, turmeric, cardamom, black pepper, and nutmeg which have become very popular across the country.

High piperine- and oleoresin-containing varieties of black pepper (IISR Girimunda and IISR Malabar excel) which are also suitable for cultivation in high altitudes and plains have been released by ICAR-IISR. IISR Thevam and IISR Shakthi are tolerant to *Phytophthora* foot rot disease. High-yielding varieties, namely, Sreekara, Subhakara, and Panchami, and rot knot nematode-tolerant variety, namely, Pournami, are suitable for all pepper-growing locations of India.

Ginger varieties, IISR Varada and IISR Rejatha, with high essential oil and oleoresin content are suitable for growing all over India. Apart from high oil content, IISR Mahima is also resistant to nematodes (*M. incognita* and *M. javanica*).

High-yielding (Suguna, Suvarna, Sudharsana) and stable high curcumin-yielding varieties suitable for growing throughout India have been developed by ICAR-IISR (IISR Prabha, IISR Pragati, IISR Kedaram, IISR Prathibha, and IISR Alleppey Supreme).

Cardamom variety (IISR-Appangala-1) is preferred widely by the oil extraction industries, whereas IISR-Appangala-2 and IISR Vijetha are suitable for mosaic-affected areas of Karnataka. Cardamom variety IISR Avinash is a rhizome rot-tolerant variety. High-yielding and high-quality varieties of nutmeg (IISR Viswashree and IISR Keralashree) have also been developed. IISR Keralashree is the first farmer's variety developed under farmer's participatory breeding program. Cinnamon varieties (IISR Nithyashree and IISR Navashree) have high shoot regeneration capacity with high bark oil and oleoresin content (Fig. 2).

It is often difficult to meet the ever-increasing demand for planting material of these popular varieties. Hence, we have initiated granting of nonexclusive licenses from 2011 for commercial production of these varieties. Healthy disease-free planting materials are provided to the licensees, and they are entrusted with the responsibility of supplying quality planting materials without deteriorating the genetic purity. The institute is now linking their clients to these licensees who in turn meet their demand. ICAR-IISR has issued 25 plant variety licenses to the clients till date.

2. **Improved Soilless Method (Pro-tray) of Healthy Planting Material Production**

(i) ***Ginger and Turmeric***

The major diseases in ginger and turmeric are soft rot caused by *Pythium* sp. and bacterial wilt caused by *Ralstonia solanacearum*. These pathogens are both seed- and soil-borne. Infection by these pathogens can be reduced by at least 50.0% through the use of disease-free planting materials. A transplanting technique in ginger by using single bud sprouts (about 5 g) has been standardized to produce

Fig. 2 Improved varieties of spices

good-quality planting material with reduced cost (Fig. 3). The yield level of ginger transplants is on par with conventional planting system. The technique involves raising transplants from single sprout seed rhizomes in the pro-tray and planting in the field after 30–40 days. The advantages of this technology are production of healthy planting materials and reduction in seed rhizome quantity and eventually reduced cost on seeds. The advantages of this method include less planting material requirement, 500–750 kg ha^{-1}, hence saving in seed cost, and only 1/3 of seed material is needed.

(ii) ***Black Pepper***

Nonavailability of healthy planting material is a serious problem in black pepper. The present technology is a solution to this. Partially composted coir pith and

Fig. 3 Production of healthy planting material of ginger and turmeric using pro-trays

Fig. 4 Black pepper seedlings in pro-trays

vermicompost enriched with *Trichoderma* is used as medium for raising seedlings in pro-trays (Fig. 4). Single-node cuttings of black pepper are planted in the trays and maintained under controlled greenhouse conditions and hardened under shade net greenhouse. The seedlings are ready for field planting within 120 days of nursery rearing. This nursery technique has enabled the production of disease-free seedlings with ease for transportation and also enhances successful establishment of vines with vigorous growth. Reduced cost of production attracts the low-income group of farmers to adopt this technology.

3. **Vertical Column Method for Quality Black Pepper**

The continuous demand for quality planting material created a novel idea of producing orthotrope on vertical 2 m column having 1 foot diameter made with plastic-coated welded wire mesh (size 4 cm) filled with composted pasteurized coir pith compost and vermicompost at 3:1 ratio fortified with biocontrol agent *Trichoderma harzianum* in polyhouse. Eight to ten cuttings are allowed to trail (Fig. 5a) on the column, and it would take 4 months to reach the top and produce

a b

Fig. 5 (**a**) Black pepper on vertical column (**b**) lateral branch

more than 20 nodes. Each vine invariably produces lateral reproductive branches within 3 months' time at 12th–15th node. The top five to seven nodes have lateral branches also (Fig. 5b). The top five nodes can be used as orthotropic shoots as is done in Malaysia and Indonesia to induce fruiting laterals from the base. In 4–5 months' time, on an average, 150 single nodes (15 cuttings per vine × 10 vines around the vertical column) per column, 1 or 2 laterals, and 10 top shoots can be harvested. The advantage of vertical column method is one can get three types of cuttings, viz., normal single-node cutting, laterals, and top shoots.

4. **Novel and Smart Delivery of Agriculturally Important Microorganisms Through Encapsulation**

Any agriculturally important microorganism like N fixers and P solubilizers can now be encapsulated and delivered to crops for enhanced growth promotion, soil nutrient solubilization/mobilization, and biocontrol. This is the world's first encapsulation technology for smart delivery of beneficial microorganisms to crops and is expected to replace the conventional talc and liquid biofertilizer formulations. At IISR, several PGPR were successfully encapsulated and delivered for growth promotion and disease control in ginger and black pepper (Fig. 6a–c). The encapsulation process is simple, does not require sophisticated equipment, and comes at low investment. Other advantages include reduced cost and easy handling and transport, no harmful by-products, less requirement of inorganic and inert material, storage at normal temperature, and, more importantly, enhanced shelf life. Besides, this encapsulation technique can be used to deliver all kinds of agriculturally important microorganisms, viz., N fixers, nutrient solubilizers/mobilizers, plant growth-promoting rhizobacteria (PGPR), *Trichoderma*, etc. Patent for this delivery process has been filed, and the technology has been commercialized by providing nonexclusive licenses to private companies.

5. **Crop-Specific Micronutrient Formulations for Major Spices**

Majority of soils in the spice-growing areas are encountering fertility issues due to acidity, nutrient imbalances, and deficiencies of secondary and micronutrients that

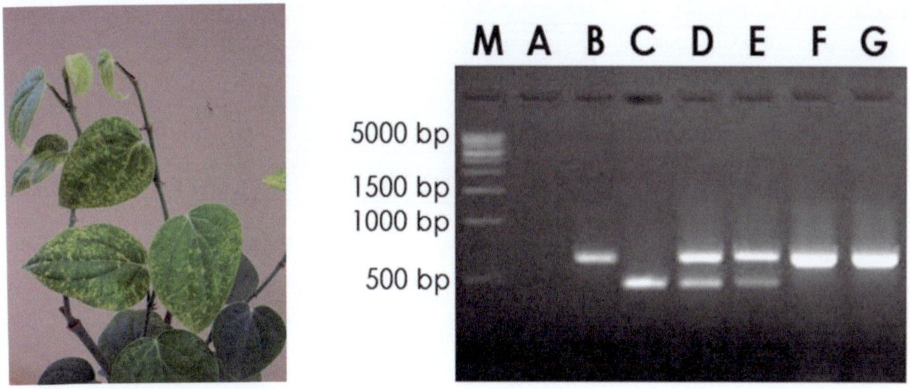

a b

Fig. 6 (**a**) Gelatine capsules containing PGPR *Bacillus amyloliquefaciens* IISR GRB 35. (**b**) Trichocap and Powercap, the two branded products of *Trichoderma harzianum* marketed by our licensee M/s Codagu Agritech, Kushalnagar, Codagu, Karnataka. (**c**) Rhizocap (encapsulated *Rhizobium* sp.), Acetocap (encapsulated *Gluconacetobacter* sp.) and Azotocap (encapsulated *Azotobacter* sp.) marketed by M/s SRT Agro Science, Chhattisgarh

a b

Fig. 7 Designer micronutrient formulations for spices

become yield limiting. Besides crop-specific, soil pH-based micronutrient mixtures for foliar application in black pepper, cardamom, ginger, and turmeric crops which guarantees 10–25% increase in yield and quality have also been developed. An innate advantage of these mixtures is that they can also be used in organic agriculture and therefore are benign and environmentally friendly. The technology comes at very low cost and hence is very farmer-friendly. The micronutrient technologies have been licensed to several entrepreneurs for large-scale production and commercialization (Fig. 7).

Fig. 8 (a) Virus-infected black pepper. (b) PCR result

6. **Pre-monsoon Irrigation for Enhanced Pepper Productivity**

Irrigation of black pepper vines around the basin from March to May at 50–80 L/vine at an interval of 15 days can markedly enhance spike length, number of spikes, oleoresin content, and berry yield. This technology promotes uniform spike initiation and reduces the spike shedding due to late monsoon and guarantees good crop.

7. **Diagnostics for Virus Infecting Spices**

Black pepper is infected by two viruses (*Cucumber mosaic virus* and *Piper yellow mottle virus*) (Fig. 8a), whereas cardamom is infected by two viruses (*Cardamom mosaic virus* and *Banana bract mosaic virus*) which are systemic in nature. Once infected, viruses cannot be eradicated by any means including chemicals. Hence, it is advisable to use certified virus-free material for planting. Loop-mediated isothermal amplification (LAMP) and real-time LAMP-based assays were also developed for quick and sensitive detection of virus diseases of black pepper and cardamom. The technology can be used for certification of mother plants/planting materials of black pepper for freedom from viruses (Fig. 8b).

A strain-specific and sensitive technique based on real-time loop-mediated isothermal amplification (real-time LAMP) was developed for detecting race 4 strain of *Ralstonia solanacearum* causing bacterial wilt in ginger. The method can be used to index both soil, water, and seed rhizomes. The time taken for detection is only 3–4 h, and the detection limit is 10^3 CFU/g of soil or rhizomes. The technology can be easily adopted in field for pathogen-free site selection as well as selecting disease-free seed materials for planting.

8. Plant Health Management Technologies

(i) *Trichoderma harzianum*: A Biocontrol Agent Against *Phytophthora*

The production of black pepper *Piper nigrum* is hampered by *Phytophthora* foot rot caused by *Phytophthora capsici* not only in India but also in other black pepper-growing countries. Talc-based bioformulation based on *Trichoderma harzianum* can be used successfully to manage *Phytophthora*. It can be used in integrated pest management as well as under organic farming system in crops like black pepper, ginger, cardamom, and turmeric. There is a great demand for the product, and IISR has already issued several licenses for its commercial production (Fig. 9).

(ii) *Pochonia chlamydosporia*: A Biocontrol Agent Against Nematodes

Plant parasitic nematodes, especially root knot nematodes (*Meloidogyne* spp.), are widely prevalent in black pepper gardens of South India and cause significant damage to the plants. Currently they are managed through application of nematicides like phorate and carbofuran. Biological control of root knot nematodes, therefore, is highly relevant in this context. *Pochonia chlamydosporia*, a known nematode biocontrol agent, is a facultative nematode parasite. It proliferates in the rhizosphere, colonizes the egg masses of root knot nematodes, and parasitizes their eggs and sedentary females. The technology is ready for transfer and commercialization to potential entrepreneurs as there is an upcoming demand for *Pochonia* in the event of ban on many popular nematicides (Fig. 10).

(iii) *PGPR Formulations for Spices*

Ginger: This consists of PGPR specific to ginger (*Bacillus amyloliquefaciens*). The major advantages observed are enhanced nutrient mobilization and nutrient use efficiency, increased growth, yield, and tolerance to soft rot disease.

Black pepper consortium: The consortium is a combination of three PGPR specific to black pepper, namely, *Micrococcus luteus*, *Enterobacter aerogenes*, and *Micrococcus* sp. It is ecologically safe, increases growth and yield, and enhances nutrient mobilization and efficiency in black pepper. It can be applied both in black pepper nurseries and under field condition as soil drench.

(iv) *Seed Coating Using PGPR*

PGPR technology is a novel process of coating efficient strains of PGPR on seeds. The components consist of live PGPR, inert material, and a binding agent. The process is done at a particular temperature which is congenial for the organisms to survive, and the coated seeds can be stored at the room temperature. Seed spices such as coriander (*Coriandrum sativum* L.), cumin (*Cuminum cyminum* L.), fennel (*Foeniculum vulgare* M.), and fenugreek (*Trigonella foenum-graecum* L.) cultivated predominantly in states of Rajasthan and Gujarat have major constraints like low

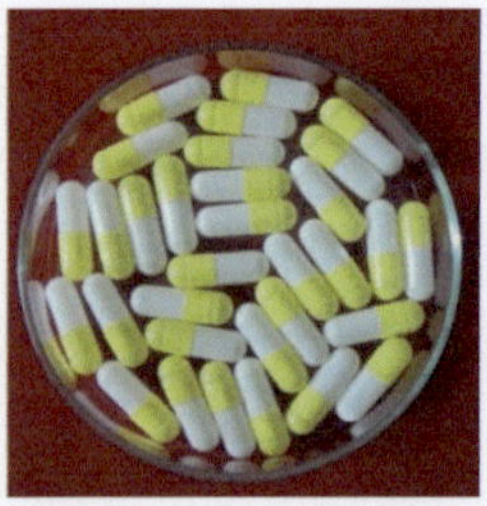

(a) Gelatine capsules containing PGPR *Bacillus amyloliquefaciens* IISR GRB 35

(b) Trichocap and Powercap, the two branded products of *Trichoderma harzianum* marketed by our licensee M/s Codagu Agritech, Kushalnagar, Codagu, Karnataka

(c) Rhizocap (encapsulated *Rhizobium* sp), Acetocap (encapsulated *Glucanacetobacter* sp) and Azotocap (encapsulated *Azotobacter* sp) marketed by M/s SRT Agroscience, Chattisgarh

Fig. 9 Talc-based bioformulation of *T. harzianum* of our licensees

germination, slow initial growth, and high susceptibility to diseases and frost. PGPR are a wide range of root-colonizing bacteria with the capacity to enhance plant growth by increasing seed emergence and crop yield. It was observed that seeds coated with PGPR exhibited longer shelf life and germination and remained intact

Fig. 10 *Pochonia chlamydosporia*

Fig. 11 (**a**) Uncoated and coated seeds of coriander, (**b**) uncoated and coated seeds of cumin after 1 year

even after 1 year of storage (Fig. 11). The technology has wide applicability and can be extended to vegetable seeds imparting the appropriate crop-specific bioagent.

(v) ***Green Technologies for Spice Cultivation***

Nutrient management plans for spices have been standardized for organic farming systems, and organic packages have been developed for black pepper, ginger, and turmeric integrating composts, oil cakes, biofertilizers/PGPRs, and biocontrol agents. In addition, an entomopathogenic fungus, *Lecanicillium psalliotae,* effective in controlling the cardamom thrips, was potentially identified and evaluated at different agroclimatic conditions in Kerala and Karnataka. Two new species of entomopathogenic nematodes viz., *Oscheius gingeri* sp. n. and *Steinernema ramanai* sp. n., are identified as potent biocontrol agent against shoot borer

Conogethes punctiferalis infesting ginger and turmeric. A new species of group I tetrahedral-shaped multiple nucleopolyhedrovirus (NPV) belonging to the genus *Alphabaculovirus* of family *Baculoviridae*, infecting *Spilarctia obliqua*, a polyphagous pest of ginger, turmeric, and other crops, was also identified as potential bioagent.

9. **DNA Barcoding Technique to Detect Adulterant in Spice Products**

Adulterants of any nature in exported commodities also adversely affect the legendary fame of Indian spices and thereby hamper the nation's prestige. DNA barcoding has been put into use to detect the plant-based adulterants in traded spices such as black pepper powder, cinnamon, nutmeg, and turmeric at ICAR-IISR. Though many vegetative adulterants such as papaya seed, wild *Piper* species, etc. are reported as adulterants in traded black pepper, DNA barcoding method could detect chile as an adulterant in traded black pepper for the first time. Of late, *C. verum* barks are adulterated with a cheaper and less aromatic bark of the morphologically similar *C. cassia* (syn. *C. aromaticum*) having a bitter and burning flavor. Our barcoding technique could detect the presence of *C. cassia* adulteration in commercial samples of true cinnamon. Similarly in turmeric, we could detect the presences of *Curcuma zedoaria* and cassava starch as adulterants in branded market samples of turmeric powder. We can detect the presence of biological adulterants like weed seeds of Lantana camera resembling pepper berries.

10. **Processing and Value Addition**

- A simple technique of hormone treatment was developed to split open nutmeg fruits without exposure to soil to prevent aflatoxin contamination in nutmeg. The methodology involves harvesting physiologically mature fruits and treating with 500 ppm ethrel (2-chloroethylphosphonic acid) for 10 min and storing, making them to 90–100% split in 18–20 h.
- A handheld electronic nose was modified with suitable sensor array for determining quality of cardamom to be graded into low (<4.0%), medium (4.0–6.0%), and high (>6%) based on essential oil content.
- Antioxidant activity of spices was tested for its nutraceutical potentials. Methanol extract of Malabar Excel was found to be highest for all the assays followed by *P. colubrinum*. Hexane extract of *P. colubrinum* showed high cytotoxicity under in vitro on cervical cancer cell line CaSki by MTT assay.

11. **Spice Processing Through Entrepreneurship Development**

ICAR-IISR has set up a business planning and development (BPD) unit, a business incubation center designed to promote entrepreneurs and equip them into profitable business ventures. The BPD unit is presently giving more impetus to entrepreneurship development and commercialization with respect to technologies developed by IISR. The BPD unit has facilitated the nonexclusive licensing and commercialization of the designer micronutrient formulations developed for black pepper, ginger, and turmeric.

Fig. 12 Products of our spice processing clients

The high-end spice processing facility established at ICAR-IISR Farm, Peruvannamuzhi, is compliant with the national and international quality requirements. The unit is equipped with state-of-the-art facilities for cleaning and grading black pepper and production of spice powders. This unit will not only cater the needs of the farmers in the spice-growing belt where it is situated but will also serve as a model unit for the benefit of spice growers and entrepreneurs all over the world. ICAR-IISR periodically organizes entrepreneurship development program (EDP) for the stakeholders to identify suitable entrepreneurs for steering forward the operations of the processing facility (Fig. 12).

New Perspectives in Spice Research

The estimated growth rate for spice demand in the world is around 3.2%, which is just above the population growth rate. The forecasted population increase is up to 1619 million in 2050 with increased GDP and per capita food spending. The per capita demand for spices is expected to increase many fold by 2050. The projected per capita demand for major spices like black pepper, cardamom, ginger, and turmeric is estimated to be about 148 g, 53 g, 1.22 kg, and 1.63 kg, respectively. With this increase, production levels to meet the local and global demand are estimated to increase by 2.7–5.7-folds from the present levels. Therefore, we need to continuously strive to increase spice productivity by enhancing input use efficiency and reducing post-harvest losses with an eye on reducing the cost of production.

Overall, the main researchable areas in spices should encompass:

- Conservation of genetic resources, bar coding, and crop improvement using cutting-edge technologies and science of *omics*

- Increasing productivity of spices through:
  - Quality planting material production and supply
  - Productivity enhancement through better input management/precision farming systems
  - Ideotype development for quality and climate resilience
  - Bio-risk management
  - Reduce labor shortage by inventing new methods of harvesting black pepper
  - Protected cultivation of spices for availability in the season
  - Enhancing the skill set of spice crops to tolerate/circumvent climate change
- New market-oriented technologies for secondary agriculture and value addition
- Exploiting the potential of spices as nutraceuticals
- Effective transfer of technologies to target group

The focus will be on newer niches where the potential for spice cultivation is immense like nontraditional areas in the Northeastern Hill Region (NEHR) where opportunities and skill sets for the future exist. In addition, research programs are to be intensified in the area of high-value compounds in spices for possible drug formulations and to develop nutrigenomics and pharmacogenomics as a composite technology in the area of bioactive compounds. Nurturing and improving sound techniques on precision farming, protected cultivation, and urban horticulture can help in surmounting the challenges posed by other growing countries. A well-reasoned and cohesive application of cutting-edge research, institutional support for development, and creative policy initiatives can ensure a vibrant spice sector in the country.

References

Boul SW (1994) Soil, society's yoke to the Earth. In: Proceedings XV of the ISSS Symposium, Acapulco, Mexico, July 10–16, vol 1, pp 89–104

Bray RH, Kurtz LT (1945) Determination of total organic and available form of phosphorus in soils. Soil Sci 59:39–45

Cope JT, Evans CE (1985) Soil testing. Adv Soil Sci 1:201–228

De Wit CT (1992) Resource use efficiency in agriculture. Agric Syst 40:125–151

Fresco LO, Kroonenberg SB (1992) Time and spatial scales in ecological sustainability. Land Use Policy 1992:155–167

Frissel MJ (ed) (1978) Cycling of mineral nutrients in agricultural ecosystems. Elsevier, Amsterdam

Keating M (1993) The Earth Summit's agenda for change: a plain language version of Agenda 21 and other Rio Agreements. Centre for our Common Future, Geneva

Morgan MF (1935) A universal soil testing system. Conn Agric Exp Stn Bull 372

Nair KPP (1984) Towards a better approach to soil testing based on the buffer power con- cept. In: Pierre-Martin P (ed) Proceedings of the sixth international colloquium for the optimization of plant nutrition, vol 4, September 2–8. Montpellier, France, pp 1221–1228

Nair KPP (1996) The buffering power of plant nutrients and effects on availability. Adv Agron 57:237–287

Nair KPP (2002) "The nutrient buffer power concept" – a revolutionary soil management technique for sustainable agriculture. Invitational research paper presented at the 17th World Soil Science Congress, Bangkok, August 2002

Nair KPP (2013) The buffer power concept and its relevance in African and Asian soils. Adv Agron 121:447–516

Nair KPP (2019) Combating global warming. The role of crop wild relatives for food security springer climate, pp 120

Robinson LR, Johnston NP, Hill JM (1981) Small-scale agriculture, an untapped giant. Benson Inst 4(2):4–7

Sanchez PA (1994) Tropical soil fertility research: towards the second paradigm. In: Proceedings of the XV ISSS Symposium, Acapulco, Mexico, July 10–16, vol 1, pp 89–104

Sharpley ANI, Pierzynki G, Sims JT (1993) Innovative soil phosphorus availability indices: assessing inorganic phosphorus. In: Agronomy Abstracts. American Society of Agronomy, Madison, WI

Stammer LB (1992) Study finds serious harm to 10 per cent of world's best soil. Los Angeles Times (quotations from World Resources Institute of U.N. Environmental Program)

K. P. Nair, *The Geography of Black Pepper (*Piper nigrum*)*,
https://doi.org/10.1007/978-3-030-52865-2

Whitmore AP, van Noordwijk M (1994) Bridging the gaps between environmentally acceptable and agronomically desirable nutrient supply. In: Glen D (ed) Proceedings of the Long Ashton Symposium: agriculture in the 21st century
Yates RA, Kiss A (1992) Using and sustaining Africa's soils. In: Agricultural and rural development series no. 6. World Bank, Washington, DC

Bibliography

Burger WC (1972) The evolutionary trends in the Central American species of *Piper*. Brittonia 24:356–362
De Waard PWF (1979) Yellow disease complex in black pepper on the island of Bangka, Indonesia. J Plant Crops 7:42–49
Gamble JS (1925) Flora of the Presidency of Madras. In: Bot Sur India, vol II, Calcutta (Repr)
Sadanandan AK, Hamza S (1996) Studies on nutritional requirement of bush pepper for yield and quality (Abs.). In: Proceedings of the PLACROSYM-XII, RRII, Kottayam, Kerala State, India, p 48
von Humboldt A, Bonpland A, Kunth C (1815) Cited by Burger (1972)

MIX
Papier aus verantwortungsvollen Quellen
Paper from responsible sources
FSC® C105338

If you have any concerns about our products,
you can contact us on
ProductSafety@springernature.com

In case Publisher is established outside the EU,
the EU authorized representative is:
Springer Nature Customer Service Center GmbH
Europaplatz 3, 69115 Heidelberg, Germany

Printed by Libri Plureos GmbH
in Hamburg, Germany